Nonlinear Algebra in an ACORN

With Applications to Deep Learning

Nonlinear Algebra in an ACORN

With Applications to Deep Learning

Martin J. Lee

Distinguished Visiting Scholar with mediaX at Stanford University,
Affiliate Program of H-STAR, the Human Sciences and
Technologies Advanced Research Institute, Stanford, CA, USA

Ken Tsang

Beijing Normal University–Hong Kong Baptist University
United International College, Zhuhai, China

World Scientific

NEW JERSEY · LONDON · SINGAPORE · BEIJING · SHANGHAI · HONG KONG · TAIPEI · CHENNAI · TOKYO

Published by

World Scientific Publishing Co. Pte. Ltd.

5 Toh Tuck Link, Singapore 596224

USA office: 27 Warren Street, Suite 401-402, Hackensack, NJ 07601

UK office: 57 Shelton Street, Covent Garden, London WC2H 9HE

British Library Cataloguing-in-Publication Data

A catalogue record for this book is available from the British Library.

NONLINEAR ALGEBRA IN AN ACORN
With Applications to Deep Learning

ISBN 978-981-3271-51-7

For any available supplementary material, please visit
https://www.worldscientific.com/worldscibooks/10.1142/11022#t=suppl

Printed in Singapore

Martin J. Lee would like to dedicate this book to his mother, who demonstrated her firm belief that the best possible education would be the best endowment to her only son by sending him to the most prestigious *Pui Ching* Middle School since kindergartens to high school. In his 9th grade in *Hong Kong Pui Ching*, Martin became interested in Algebra and became the best student in the mathematics class under the tutelage of Mr Lum Ying Ho who also taught Daniel Chee Tsui a year later.

Foreword

Normally, you would expect the authors to ask a world class mathematician to write a foreword to this volume. However, they gave the honor to me, a noted software engineer and innovator. Besides being a friend, I believe the reason why they did this is because I can sense the applicability of this work in optimization of non-linear objective function and the ACORN approach especially as aided by machine learning and understanding techniques to such fields in which I have practical experience such as industrial automation, finance, marketing, manufacturing, oil exploration, and telecommunications. This practicum becomes extremely important to engineers like myself as we seek to solve complex problems that involve many variables operating in n-dimensional mathematical space like those involving determination of where to move, control and/or use resources like those needed to produce goods, operate machines and/or determine the best way to make investments. Our goal in such a quest is to develop a best rather than a singular answer because there may be many solutions available.

Let us look at an example. Currently, oil companies explore for oil offshore by conducting seismic surveys. They identify potential geographical areas based on geology and many other variables and then blast underwater to gather data that allows them to map the ocean's floor. Such mapping applications require them to run very sophisticated software on multiple supercomputers in parallel for months at a time to develop models of the ocean's floor. Engineers often fight for computer time because developing such models is so compute-time extensive. As you can imagine, this is a very expensive process whose output is justified by discovery. A variety of mathematical routines like those described in this book can be used to increase such probabilities of discovery.

Let us look at a second example. In the future, thinking machines will process vast amounts of data as they seek solutions using training sets to home in on optimal solutions. Such machines will tackle just about every problem known to man as they determine how best to operate when dealing with many variables and consequences as they make decisions. The algorithms they will use will go well beyond those simple neural networks discussed in this pamphlet as will the complexity of the issues they will face. For example, decisions that involve safety and security will have to be subjected to exception conditions to determine whether or not they are suitable especially in the work place, home and environments in which they operate.

As you can see, mathematics like that discussed within this document has a great deal of applicability in the real world. Hopefully, highlighting this use will give you some additional context for review and potential use of this work outside the realm of particle accelerators and other science-based applications.

Donald J. Reifer
Software Engineer, Innovator and Businessman

Preface

This book was written as a first course for interested students and researchers to learn a new approach to optimization of nonlinear objective functions in high dimensional space.

This method has its origin in the Stanford Linear Accelerator Center (SLAC) in the 1960s when the first author (MJL) was working as an accelerator designer, initially a graduate student under Professor Burton Richter and later as research staff and leader of the accelerator design group in SLAC.

Accelerator design was a serious business in the 1960s and 70s, and was crucial to the discovery of the J/ψ meson in SLAC announced on 11 November 1974. Two teams of experimental physicists, one headed by Samuel Ting of MIT using the accelerator at the Brookhaven National Laboratory and the other headed by Burton Richter in SLAC, were in a dead heat competition to claim the discoverer of this illusive particle that would bring new order to the "elementary particle" zoo at the time. Both of them were awarded the Nobel Prize in Physics in 1976. Since then, the SLAC accelerators have produced two more Nobel Prizes in Physics for the quark structure inside protons and neutrons (1990) and the tau lepton (1995).

MJL's expertise is in the frontier of accelerator design, using numerical algorithms to adjust accelerator beam-line parameters to obtain best particle beam quality and highest possible energy as well as beam intensity. By playing with the hardware parameters, such as the number of field coils, magnetic dipole, quadrupole and octopole strengths, accelerator designer hopes to achieve the highest possible beam intensity and best beam quality that is consistent with the energy constraint and other experimental considerations.

Mathematically, it is an exercise of optimizing an objective function, which represents the beam intensity and/or beam quality, of multivariables, which are the accelerator beam-line parameters. The objective function in this case is usually a nonlinear function of hundreds, sometimes up to thousands, of variables, due to the fact that the main accelerator of SLAC is 3.2 kilometers (2mi) long—the longest linear accelerator in the world and constituted of millions of hardware components. An optimization problem in such a high dimensional space is indeed very challenging.

After years of hard work and experience, trying various well known and not so well known numerical recipes, in public domain or non-public domain, MJL came up with his own approach to optimization and tested it in the SLAC accelerator design over the ensuing forty years. This book is the result of such experience and an account of the new approach to optimization.

Below is a brief elucidation of the central problem focused in this book, in the form of questions and answers.

- **What are we after?**
  Finding solutions that satisfy a known relationship, expressed in terms of a set of equations, linear or otherwise, is a central problem of mathematics since ancient time. While finding the solution of a linear equation system either analytically or numerically, the subject matter in linear algebra, is straightforward, nonlinear equation systems of various forms are far more challenging to deal with. Even for a general polynomial equation with single variable, there is no analytic solution of its roots beyond quartic.

  Most often, even the existence of the root is not sure, not to mention to locate where it is and obtain an analytic or approximated numerical solution. This is exemplified by the famous Riemann hypothesis, which is listed as one of 23 unsolved important problems by the renowned 20th century mathematician David Hilbert and still currently one of the Clay Mathematics Institute's unsolved Millennium Prize Problems.

- **Why do we care?**
  Despite all these difficulties, root-finding for nonlinear equations is extremely important in many practical engineering and scientific applications, for example the eigenvalue problem, with many

critical modern technologies depending on it.

Furthermore, root-finding is essential to another important branch of mathematics, optimization, i.e. the art of locating maxima or minima of a real function. We call it an art because there is no sure way to arrive at the answer routinely and efficiently. The method of solution is to a large extent problem dependent. Generally speaking, the goal of optimization is to seek the "best available" values of some objective function, which in this book we refer to as an objective function, within a given domain (of input parameters). This turns out to have extremely important applications in many disciplines of modern human knowledge, including Agriculture, Banking, Control, Engineering, Finance, Manufacturing, Marketing, and Science. To quote just a few examples: In physics, almost all physical systems in equilibrium states are the results of either minimizing some sort of internal energies, or maximizing the entropies of the systems. In economics, the day-to-day problems facing policy-makers and corporation managers involve maximizing efficiencies or some kind of "utilities" (such as profit) or minimizing costs (or waste) under various situations.

At this point, it is important to recognize that finding the maximum of an objective function f is equivalent to locating the minimum of $-f$. Hence, without loss of generality and to simplify our presentation, in the remainder of this book, we will focus our attention and discussion on methods to minimize an arbitrary objective function, f. The minimization algorithm we present later on in this book can be easily modified to serve as a maximization algorithm. The term "minimizing an objective function", henceforth, is equivalent to "optimizing an objective function".

- **What do we offer in this book?**
  A simple algorithm for solving a set of nonlinear equations by matrix algebra has been discovered resulting from the forty years' experience of MJL in SLAC-first by transforming them into an equivalent matrix equation and then finding the solution analytically in terms of the inverse matrix of this equation. With this newly developed algorithm, which is aptly named **ACORN**, an acronym from "**A**daptive **C**onstrained **O**ptimal **R**obust **N**onlinear", it is possible to minimize the objective function

without computing its derivatives. The convergence of this non-linear analytic iterative scheme requires the choice of two control parameters (regardless of the dimensionality of the problem), judiciously chosen to collapse a high dimensional search to just a two dimensional search. This book will present details of the ACORN algorithm and how can it be deployed to solve large scale nonlinear equations with an innovative approach ACORN Magic [Minimization algorithms gathered in a cloud].

The ultimate motivation of this work is ACORN's application to optimization. To be sure, solving nonlinear equation has many applications in scientific and engineering problems. In recent years, with the advances in Big-Data technology, optimization becomes an even more powerful tool in knowledge discovery.

In problems of equation solving, the number of unknowns is equal to the number of equations of the system we are working on. In other words, the number of constraints or relationships we need between the unknowns has to be exactly equal to the number of unknowns we want to determine. When the constraints we know are less than the number of unknowns we try to uncover, the problem is under-determined and has no definite answer. Simply put, if we do not have complete information about the system we are going to study, equation solving algorithms will not help. Sadly, this is the situation we found ourselves facing in most scientific and engineering applications. The world is far from ideal.

Fortunately, we can find a way out of this quandary by trying to fit all the data we have to the best model that is consistent with our incomplete information. In statistics, this is known as "regression analysis", which is basically a procedure for optimizing (i.e. minimizing) the sum squared error between the actual data and the model prediction, by adjusting a set of model parameters. The result is the best model that is consistent with all the data we have. Even though our information may not be complete, we can still make prediction to the best of knowledge we have. Furthermore, with the large amount of data we have in this "Big-Data" era, the probability that our prediction is correct becomes higher and higher.

- **How can ACORN help in Deep-Learning?**

Modern day 'machine' (more appropriately, 'statistical') learning algorithms, is an out-growth from the work by pioneers in artificial intelligence (AI) during the 1950s. It began as an attempt to study how to construct algorithms that make data-driven predictions or decisions instead of making decisions by following static pre-programmed instructions. Today, machine learning has gained wide acceptance in pattern (speech and image) recognition, computer vision, as well as teaching machine to acquire the intelligence to play chess against human.

Artificial neural network (ANN) is by far the most popular and successful approach to machine learning inspired by the biological neural networks that constitute the human brain. Such network is based on a collection of connected artificial neurons (analogous to neurons in a biological brain) stacked up in a layer that can transmit signals (i.e. data) through connections (i.e. synapses) to neurons of the next layer. Neurons and synapses in each layer are characterized by weights that are determined by the training data and varied as learning proceeds. The entire learning process, which mimics the human learning process, is mathematically analogous to a statistical regression, with the objective function determined by the specific architecture of the ANN. The final outcome of the learning process is the optimized set of synaptic weights which minimized the sum square errors between the output of the ANN and the actual data it is supposed to predict.

The deep neural network (DNN) that has caught much public attention recently is an ANN with a large number of hidden layers between the input and output layers that can be used to model very complex non-linear relationships. Again DNNs are modelled after the hierarchical abstraction in human thinking, and are typically architected as feedforward networks in which data flows from the input layer to the output layer. The advances in DNN technology are credited as the force behind recent rapid progresses and commercialization in computer vision, and automatic speech recognition.

The news that *AlphaGO*, a software developed based on DNN, defeated top human GO masters in 2016 and again in 2017 is a demonstration of the awesome power of AI in general and DNN in particular. We do not have to teach *AlphaGO* how to play GO.

Somehow, *AlphaGO* is able to learn the game by itself if it is given sufficient data to learn.

The knowledge we get from machine learning is not 100% exact because of its statistical nature, however, it is good enough for us under most situations. *AlphaGO* is a good example of the usefulness of AI through DNN.

Despite all these glories, however, few in the general public realizes that inside all machine learning algorithms (including DNN) there are optimization engines driving the process of searching for the best set of model parameters that is consistent with the data fed into the learning process.

All machine learning processes are data-driven and rely on algorithms that can process vast amount of learning data automatically to extract a definite trend or conclusion. Deep learning, with a large number of hidden layers in the neural network, involves the picking of an extreme large number of network parameters (the synaptic weights) in order to minimize the objective (error) function. As we will show later in this book, optimization in high dimensional space, such as that encountered in the training of DNN, is notorious for its complexity and pitfalls. This phenomenon is commonly known as the curse of dimensionality in nonlinear data modeling.

Many issues can arise with naively trained DNNs. Two common issues are overfitting and computation time. DNNs are prone to overfitting because of the added layers of abstraction, which allow them to model rare occurrence in the training data. DNNs must also consider many training parameters, such as the size (number of layers and number of units per layer), the learning rate and initial weights. Combing through the high dimensional parameter space for optimal parameters may be infeasible due to not only the cost in time and computational resources, but also its complexity.

As all students of optimization have often been warned, there are local minima besides the global minimum. Even in one-dimensional function, finding a local minimum does not necessary mean that the global minimum is found. Global solutions are needed in most applications, but they are usually difficult to recognize and even more difficult to locate. Only for special situations, such as convex or linear programming problems, local solutions are also global solutions. General nonlinear problems, both constrained and unconstrained, may possess local solutions that are not global solutions.

Increasing the dimensionality from one to two, the 3-D terrain that represents the variation of a 2-D function begins to tax on our ability to visualize the problem. The 3-D landscape may contain peaks and valleys, as well as saddle points, a new kind of phenomenon not seem in any 2-D landscape. In N-dimensional space as N getting bigger and bigger, we lose our intuition quickly because our mind, formed with experience in a 3-D physical world, cannot visualize the convoluted geometry in higher dimensions. Moreover, the complexity of the optimization problem grows exponentially, making it difficult even with a powerful computer.

All existing optimization algorithms discussed in the literature have no idea whether the answers they finally converge to are local or global minima, even if they converge at all. Under some circumstances, the local minima are data dependent, i.e. the number of locations of local minima vary with the data sample used in the training, and hardly can be considered as the valid answer to the optimization problem at all. Hence, the ANN or DNN resulted from such training may not be reliable or satisfactory.

To advance the technology of machine learning and AI to a higher level, a new generation of optimization algorithm is needed. The widepread application of machine learning and AI so far is just the beginning. There are a large number of problems waiting to be explored by machine learning.

The ACORN algorithm and its implementation with cloud computing, ACORN Magic, as presented in the book, is the perfect choice in this kind of applications because of the fact that it is fast and robust in identifying the global minimum, and simple enough to be embedded in any type of machine learning program.

This book is organized in a straightforward manner.

In Chapter 1, we reviewed two most widely used optimization algorithms for convex object functions, i.e. the gradient descent method and the Newton's method. From these the more sophisticated Levenberg-Marquardt optimization algorithm was motivated. Then we borrowed the concepts of attractor and basin of attraction from the theory of dynamical systems to analyze the behavior optimization algorithms. Based on these concepts, we illustrate how different optimization algorithms can end up converged to different minima even starting from the same initial guess. Hence, all gradient based optimization techniques have no control of their destiny. They are never sure the answer they converged to is a global or local minimum.

In Chapter 2, we developed the ACORN algorithm in detail from the ground floor up. Effectively, this algorithm is a scheme to collapse the multi-dimensional search for global minimum to a 2-dimensional one. Fundamentally by its searching nature, the ACORN algorithm can avoid being trapped by local minima and eventually smell its way to the global minimum, regardless of the initial guess.

In Chapters 3 and 4, we applied the ACORN algorithm to simple examples as warm-ups before the more important applications to deep learning.

Chapter 5 is dedicated to deep learning. After a brief review of DNNs and how the learning process works, we gave two examples, from simple to more complicated, of how the ACORN algorithm can be applied to train DNNs. A discussion of how the parallel nature in the ACORN search algorithm can be exploited to take advantage of current advances in cloud/GPU computing technologies is presented at the end of the chapter.

Acknowledgments

The authors would like to thank Drs. Alfred Kwok, Ting Ho and Steve Huang for their numerous valuable suggestions to improve the manuscript and proof-reading, Professor Michael Saunders, Director of Systems Optimization Laboratory of Stanford University, for his stimulating discussions, Dr. Martha G. Russell, Executive Director of mediaX at Stanford University, for her constant encouragements, Dr. C. K. Wong, Director and founder of iASPECT Technologies, for his help to perform testing and validation of the algorithm.

We also thank the following friends and colleagues for their encouragements from the initial conception and during the process of writing of this book: Prof. Richard Blankenbecler, Ho-Ping Chang, Yi-Shin Chen, Ping J. Chou, William J. Corbett, Michael Dikovsky, Kun Fang, Nen-Fu Huang, Xiaobiao Huang, Kun-Han Lee, Prof. Lou-Chuang Lee, Dr. Li Jiong-Cheng, Ho Chung Liu, Henry Lui, Patrick Lui, and Juhao Wu.

Finally, we are greatly indebted to Drs. Paul Chu and Weibin Liu for the data used in Chapter 4, to Frank Ou, who contributed enormously to the examples in Chapter 5 and in Appendixes C and D, and to Zhao Chenchen for preparing the LaTeX manuscript.

Last but not least, we are deeply grateful for the constant encouragements and supports from Mrs. Emily Lee and Susan Tsang over the years since we worked together on optimization algorithms.

The writing of this book is partially supported by internal grant R201626 from BNU-HKBU-United International College, Zhuhai, China.

Contents

Chapter 1

An Overview of Optimization

1.1 Two Simple Optimization Schemes

A simple and intuitive technique to find the minimum of a function is the *"gradient descent"* method. It is simple because it is an iterative optimization algorithm involving just the first derivative. To find a local minimum of an *objective function*, one starts with an initial guess (point) and takes successive steps with step-size proportional to the negative of the gradient of the *objective function* at its current position (point). In other words, user initializes the procedure with a guess and takes a small step to slide down along the local slope to the next point where the *objective function* should be smaller if the step taken is sufficiently small. Repeat the procedure until the local minimum is reached, or stop when the maximum number of steps is exceeded.

In practice, there is one potential flaw with the gradient descent rule, because we should take smaller steps down the gradient when the gradient is large, so as not to miss the minimum. On the other hand, we would like to take larger steps at locations where the gradient is small so that the algorithm converges to a minimum faster. However, with a fixed constant of proportionality in the *gradient descent* rule, we do just the opposite of what we should.

Another issue is that the curvature of the *objective function* may not be the same in all directions in an N-dimensional problem with $N > 1$. For example, if there is a long and narrow valley in the N-dimensional surface that serves as a graphical visualization of the *objective function*, the component of the gradient in the direction that points along the base of the valley is small while the component along the valley wall is quite large. This results in a path of the successive iteration points more in the downward direction of the wall even though we should move a longer way along the base and just a small bit along the wall, in order to reach the minimum.

In the *"gradient descent"* method we assume the *objective function* is differentiable, we use its first derivatives to guide us to the optimal point.

The second common iterative numerical method to optimize an *objective function* is to apply *Newton*'s method to the first derivatives of a twice-differentiable *objective function*, in order to find the roots of the first derivative. *Newton*'s method, which can be found in any elementary collection of numerical recipes, is a popular root-finder for solving nonlinear equation. In this algorithm, the second derivative of the *objective function*, i.e. the curvature, is needed. The main advantage of this technique is its rapid convergence. However, the rate of convergence is sensitive to the initial guess, and sometimes may miss the local minimum entirely.

1.2   The Levenberg-Marquardt Optimization Algorithm

Levenberg [Levenberg (1944)] observed that the convergence rates of the simple *gradient descent* method and the *Newton* iterations are somewhat complementary to each other. To take advantage of both, he proposed an algorithm, whose update rule is a blend of the two with an adjustable relative weighting factor favoring one of the algorithms depending on the situation. In effect, the Levenberg rule follows closely with *Newton*'s method if the *objective function* is reduced after such an update. If the *objective function* goes up, then the weighting factor is changed so that the step-size is determined mostly by the *gradient descent*.

However, in Levenberg's original algorithm, the problem due to a narrow and gentle valley with steep walls in the N-dimensional surface of the *objective function* was still not yet addressed. It was due to Marquardt's insight [Marquardt (1963)] that the relative weighting factor was later modified by inclusion of the curvatures of the *objective function* (i.e. the diagonal terms of the Hessian matrix) so that the *gradient descent* step-size is reduced along the steep valley wall and increased along the narrow and gentle valley basin, since the Hessian matrix was already calculated and used in the Newtonian part of the algorithm anyway.

From that point on, the Levenberg-Marquardt (LM) minimization algorithm becomes an industrial standard. The LM method is by no means optimal and is somewhat heuristic, but it works extremely well in practice. For moderately sized models (of a few hundred parameters), the LM method is much faster than the *gradient descent* method alone. The LM algorithm is more robust than *Newton*'s method, which means that in many cases it finds a solution even if it starts so far off the final minimum that *Newton*'s method failed.

However, there are a number of drawbacks associated this kind of algorithms which makes this popular method neither satisfying nor reliable. A few of them can be listed here:

(1) Computation of first and second derivatives is required, so that it is time consuming, especially in high dimension; and the derivatives may not exist;

(2) Matrix inversion is needed as part of the update, so that the cost of update becomes prohibitive after the model size increases to a few thousand parameters;

(3) At most one local minimum can be located each time, no exhaustive search of all the minima and hence no knowledge of the global minimum is provided;

(4) In cases when there are multi-minima within the domain of study, gradient descent in general and Levenberg-Marquardt in particular would likely be trapped in a local minimum and not able to locate the global minimum.

1.3   General Characteristics of Optimization Algorithms

Starting from these two archaic approaches to optimization, a plethora of algorithms have been developed in the last fifty years for optimization of smooth objective functions under various constraints. We refer serious readers to an extensive text *Numerical Optimization* by Nocedal and Wright [Nocedal and Wright (2006)] for this purpose.

At this point we would like to take note on the general characteristics of any numerical optimization algorithm in N-dimensional space, because this will help our readers to achieve a deeper level of understanding in the chapters below. Both the optimization and root-finding are mathematical problems that share many features in common, as illustrated in the discussions above how Levenberg and Marquardt were motivated to come up with the LM algorithm.

Solving the optimization or root-finding problem requires the identification of a particular object (i.e. point) in a space of possibilities (i.e. the N-dimensional space of interest) that satisfies certain requirements. In the former it is to search for the object that associates with a minimum in the objective function, and in the latter with a zero value in the objective function. This space that we are searching, can be, and usually is, enormous due to its high dimensionality.

Consequently, the optimization or root-finding problems that are to be solved this way can be seen as search problems. Any numerical algorithm to serve this goal is just a recipe of searching routine that hopefully will lead us to the object we are after in this vast searching space.

This kind of searching routine starts from an initial point in the searching space, usually supplied by the user using an educated guess. From that point on, the searching routine will take control of its own destiny by entering into an iterative loop that generates successive answers to the problem at hand. Each answer is supposedly a bit better than the one from previous iteration. The set of rules used to move from current iteration to the next distinguishes one algorithm from another. The process ends when either the maximum number of iteration is reached, or the absolute difference between the last two successive answers is so small that they are regarded as practically the same. If the latter case is true, the searching routine is converging to a final answer for our problem. Otherwise, the searching routine is not converging within the number of iterations specified by the user.

When the searching routine fails to converge to an answer, we can either increase the maximum number of iterations and try again, or give up and start all over with a different algorithm. The reality is there is no guarantee that any of the optimization algorithm presented in literature will definitely lead to the solution of our specific problem. The arguments that motivate an algorithm are usually heuristic and cannot be confused with a mathematical proof of convergence.

Most optimization algorithms require the specification of certain parameters that control the rate of convergence, besides the initial guess. In the case of the LM algorithm, it is the parameter that controls the switching between the *Newton*'s method and gradient descent method. To assure good convergence, the control-parameter has to be tuned from time to time, depending on the problem of interest.

Regardless all the detail inner working of an algorithm, a user ultimately judges the usefulness of an algorithm by the following properties, as suggested by Nocedal and Wright [Nocedal and Wright (2006)]:

(1) Robustness: It should converge to a solution on a variety of problems, for all reasonable values of initial guesses.

(2) Efficiency: It should not require excessive computational time (i.e. algorithm should converge within a reasonable number of iterations) or storage.
(3) Accuracy: It should not be overly sensitive to the arithmetic rounding errors that occur when the algorithm is implemented on a computer.

1.4   Optimization Algorithms and Dynamical Systems

From the previous discussion above, it becomes obvious that an optimization algorithm has a strong resemblance to a dynamical system. Mathematically, a *dynamical system* is defined as a system with a mapping that governs the time evolution of a point in the N-dimensional space, which is referred to as the state space because it represents the totality of all possible states of the *dynamical system*. Often, the mapping is deterministic, that is, given a fixed time-step, only one possible future state follows from the current state.

A physical example of a continuous *dynamical system* is the orbital motion of Earth around the Sun. At any instance in time the Earth moves from its previous location to a new location given by its Cartesian coordinates x, y, z with new velocities V_x, V_y, and V_z. The *state space* in this example is the 6-dimensional space of (x, y, z, V_x, V_y, V_z), and its orbit is a continuous path in this state space. The mapping that moves the Earth from one moment to the next in the state space is Newton's second law of motion with the gravity between the Earth and the Sun.

In an optimization algorithm, we do not have time explicitly, but we have an iterative loop that works like a clock, i.e. it moves only in forward direction. The iteration starts from an initial point in the N-dimensional space. Following the recipe of the searching routine, the point in N-dimensional space that describes the state of our optimization process is moved to a new location after each iteration, much like the Earth is moved to new locations with new velocities day after day. This shows that an optimization algorithm is indeed a garden variety of *dynamical system*. In particular, it is a discrete *dynamical system*, in which the time (i.e. the number of iterations) takes on only non-negative integer values, and the mapping in this *dynamical system* is given by the recipe inside the iterative loop.

From this point on, we can borrow the mathematical concepts and results developed for *dynamical system* in our study of optimization algorithm. In particular, the concepts of '*attractor*' and its '*basins of attraction*' are very useful in the discussion of optimization algorithm.

1.5   Attractors and Basins of Attraction

Loosely speaking, an *attractor* is a subset of the N-dimensional *state space* to which state-paths (or orbits) originating from some initial states (points in the N-dimensional space) approaches as time increases indefinitely. It is common for *dynamical systems* to have more than one attractor. For each *attractor* in a *dynamical system*, its *basin of attraction*[1] is the set of initial states leading to the long-time behavior that approaches that *attractor*.

In general, an *attractor* does not consist of just a single point. In the example above of Earth's orbital motion around the Sun, the *attractor* is a '*limit cycle*', a periodic orbit in the state space (x, y, z, V_x, V_y, V_z), if gravity follows the inverse square law.

In a discrete-time dynamical system, an *attractor* can take the form of a finite number of state points. In more exotic situations (i.e. chaotic dynamical systems), the *attractors* can be very complicated. The reader may have heard of the term '*strange attractors*', but for now they are not our concern.

Attractors associated with an optimization algorithm are the optimal points of the *objective function*. When the algorithm converges to an answer, the attractor of the algorithm is a single *fixed-point*. Otherwise, it is not a *fixed-point attractor*. There may be more than one *attractor* associated with an optimization algorithm if there are many local minima. Each local minimum of the *objective function* is an *attractor* by itself, and has its own basin of attraction, which consists of all the initial guesses that converge to the same local minimum.

Thus, a *basin of attraction* for an optimization algorithm is the set of initial values (guesses) leading to the same local minimum. Generally, the basic topological structure of *basin of attraction* can vary greatly from algorithm to algorithm. It is an important measure of the robustness of the algorithm.

An algorithm with a larger *basin of attraction* for a local minimum of interest is a more robust algorithm compared to another one with a smaller *basin of attraction* for the same local minimum, if that particular local minimum happens to be the answer you are seeking, because you have a

---

[1]See the details in http://www.scholarpedia.org/article/Basin_of_attraction

better chance to find it. It is for this reason that the study of *basin of attraction* may shed more light to the robustness of an algorithm.

1.6 Basins of Attraction for Optimization Algorithms

Let us illustrate the concept of attractor and basin of attraction with an example using a more generalized form of Rosenbrock function, which is a non-convex function cooked up by Howard H. Rosenbrock in 1960 to demonstrate how a long and deep valley with high walls can appear in 2D. The simplest form of Rosenbrock function is 2D function defined by: $F(x, y) = (a-x)^2 + b(y-x^2)^2$, where a and b are constant parameters. It has a global minimum at the point with $x = a$ and $y = a^2$, where $F(x, y) = 0$. The parameter b does not change the answer, but it affects the landscape of the 2D surface. If b is a number much larger than unity, then we have a deep valley with steep walls along the parabolic curve $y = x^2$. The global minimum is located inside this deep valley with relatively smooth basin.

In our example here, we use a slightly more complicated version:

$$F(x, y) = x^2(a - x^2)^2 + b(y - x^2)^2 + c(x^2 + y^2)^2, \qquad (1.1)$$

where a, b and c are constant parameters. We are interested in the case where $b \gg 1$ and $c \ll 1$, so that the global minimum is located exactly at $(0, 0)$ with $F(0, 0) = 0$. There are two local minima in this objective function as well. They are roughly located near $(+\sqrt{a}, a)$ and $(-\sqrt{a}, a)$, if c is much smaller than unity. Take the case of $a = 1$, $b = 5$ and $c = 0.01$, the contour plot of this objective function is shown in Fig. 1.1 for a domain defined by $-2 < x < 2$ and $-1 < y < 3$, which looks like a smiley face. We choose a moderately large value for b just enough to show the effect of the steep walls of the valley, but not so much as to strain its visibility.

All these three points, where the minima located, are the 3 fixed-point attractors for any optimization algorithm to converge upon. Figures 1.2(a) and 1.2(b) below show the basins of attraction (BOA) for the objective function of the form given in Eq. (1.1) whose contour plot shown in Fig. 1.1. Figure 1.2(a) shows the basins of attraction associated with the default algorithm (Interior-Point) of fmincon function supplied by the optimization toolbox of MATLAB, and Fig. 1.2(b) shows the same thing associated with another algorithm with the acronym: 'SQP' (Sequential Quadratic Programming). The technical details of these specific algorithms will not be discussed here as they are only tangential to our main goal, and the reader

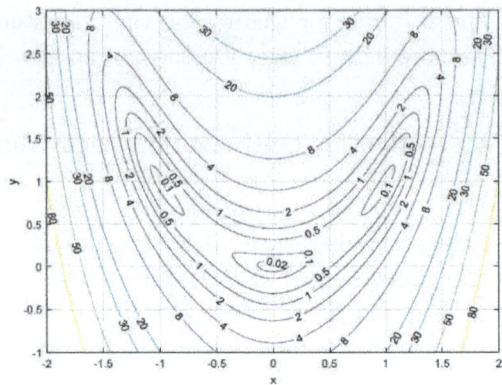

Fig. 1.1: Contour plot of an objective function of the form given in Eq. (1.1) with $a = 1$, $b = 5$ and $c = 0.01$ for a domain defined by $-2 < x < 2$ and $-1 < y < 3$. The deep valley is along the parabolic curve $y = x^2$, where the 3 minima are located. The global minimum is located at $(0, 0)$ with $F = 0$. The other 2 local minima with $F > 0$ [*actually* $= 0.0364$] are roughly near $(1, 1)$ and $(-1, 1)$ [more exactly at $(-0.971, 0.936)$ and $(0.971, 0.936)$]. All these 3 points are the fixed-point attractors for any optimization algorithm to converge at.

is asked to refer to the documentation[2] provided by MATLAB. Both of these two algorithms are discussed in detail in the textbook by Nocedal and Wright [Nocedal and Wright (2006)].

In both Figs. 1.2(a) and 1.2(b), the blue region is the BOA for the global minimum at $(0, 0)$, whereas the red and green regions are BOA for the local minima at $(-0.971, 0.936)$ and $(0.971, 0.936)$, respectively.

The first thing an observer would notice is the amazing complexity of the topology in these three regions, which inter-mingle with each other in a complicated manner, for both BOA plots in Figs. 1.2. It is quite startling that a small pocket of green (or blue) can be buried deep inside the red region when it is far away from the attractor it is supposed to converge to, and likewise for the other colors. This means that sometimes even if our initial guess is close to the solution we want, but unfortunately it is inside one of those small pockets of BOA that belongs to an attractor farther away, the algorithm would not converge to the one we anticipated. Even the chance for that to occur is not high, however, Figs. 1.2 tell us that we should not be surprised if it did occur.

---

[2]Detailed discussions of the algorithms available to `fmincon` can be found in
https://www.mathworks.com/help/optim/ug/
constrained-nonlinear-optimization-algorithms.html
and https://www.mathworks.com/help/optim/ug/choosing-the-algorithm.html

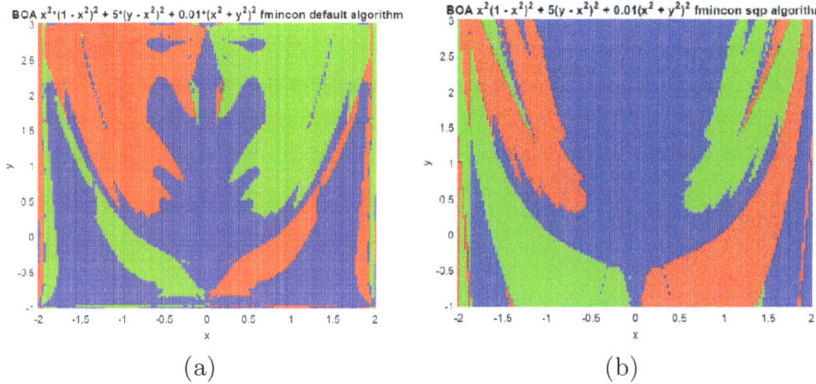

(a)   (b)

Fig. 1.2: (a) The basins of attraction (BOA) for the objective function whose contour plot is shown in Fig. 1.1, produced by using the default algorithm (`interior-point`) of `fmincon` in the optimization toolbox of MATLAB. The blue area is the BOA for the global minimum at $(0,0)$, whereas the red and green regions are BOA for the local minima at $(-0.971, 0.936)$ and $(0.971, 0.936)$. Note the complexity of the topology in these three regions, which inter-mingle with each other in an amazingly complicated manner. (b) BOA for the same objective function with same color coding, using the MATLAB 'SQP' algorithm instead. Note the big difference in their appearance compared with the BOA in (a). The Octave code for the BOA plots is listed in Appendix A.

The second observation one would get from comparing the BOA plots is BOA depends strongly on the algorithm used, as the two plots look so different. Beyond the fact that the two BOA plots appear to have same set of attractors, we cannot find any clue to suggest that the two BOA plots are associated with the same objective function. If we are interested to find the global minimum, a larger blue region in the BOA plot means a higher probability to select an initial guess that would converge to the global minimum at $(0,0)$.

The lesson one would learn from Figs. 1.2 is that not all algorithms are created equal. BOA plots may serve as a fair and useful comparison between the performances of two algorithms. In the example above, we have selected a bounded domain in 2D where most of the interesting behaviors of the objective function are located so that all points within the domain eventually converge to an attractor. However, this is not necessary true in general. Sometimes there may be points within the domain under consideration do not belong to any basin of attraction, due to the algorithm does not converge to any attractor[3]. Thus the size (area) of a basin of attraction

---

[3]An example of this is shown in `http://www.mathworks.com/matlabcentral/fileexchange/43564-basins-of-attraction?focused=3798465&tab=function`.

may serve as a quantitative measure of the robustness of the algorithm with regard to an attractor.

If by *robustness* we mean the algorithm should converge to a solution we want on most problems with reasonable values of initial guesses, then very few of current optimization algorithms can be regarded as robust.

Summary of Chapter 1

The key points that can be gleaned from this chapter include the following:

- How the gradient descent method and the Newton iteration can be combined to form the backbone of the **Levenberg-Marquardt optimization algorithm**, a work-horse for the modern optimization methodology.
- The similarity between an optimization algorithm and a dynamical system, and how the robustness of an optimization algorithm can be analyzed by studying its Basin of Attraction, a concept borrowed from the theory of dynamical systems.

Chapter 2

The ACORN Approach to Optimization

2.1 Transform Nonlinear Equations into a Matrix Equation

As we have already explained why optimizing a smooth objective function of m variables is equivalent to solving a set of m nonlinear equations with m unknown. This is because optimization can be achieved by locating the zeros of the first partial derivatives of the objective function. In the following, we shall develop a novel approach to solve a set of m nonlinear equations with m unknown. This approach is based on a transformation to rewrite the set of nonlinear equations into a matrix equation [Lee (2012)].

For a set of m equations with m unknowns $(x_i, i = 1, \cdots, m)$, it is more convenient in our approach to express them in the form shown below where the right hand side is normalized to the constant 1 (unity) and all other dependence on x_i is combined to define the functions f_i on the left-hand side. It will become apparent later why this particular form is useful to us.

The next step is to rewrite the set of nonlinear equations

$$f_i(x_1, x_2, \cdots, x_m) = 1 \tag{2.1}$$

formally as a matrix equation:

$$M\underline{X} = \underline{1} \tag{2.2}$$

where $\underline{X}$ is the transpose of the m-dimensional row vector $(x_1, x_2, \cdots, x_m)$ and $\underline{1}$ is the m-dimensional column unit vector with all components equal to unity.

The square $m \times m$ matrix M is defined in such a way that after all the trouble of matrix multiplication is carried out we are back to our starting point, i.e. recovering the original set of nonlinear equations:

$$f_i(x_1, x_2, \cdots, x_m) = 1. \tag{2.3}$$

This apparently futile exercise forms the starting point of our ACORN algorithm. The matrix M is well-defined as long as none of the components in $(x_1, x_2, \cdots, x_m)$ is equal to zero, which we will assume in this work. Attentive readers may also notice that we have introduced a new parameter 'r' in the definition of M. It turns out that this parameter 'r' will play an instrumental role in finding the roots of the set of nonlinear equations:

$$f_i(x_1, x_2, \cdots, x_m) = 1, \tag{2.4}$$

in an iterative procedure.

We leave the algebraic details above as an exercise to the reader.

$$f_i(x_1, x_2, \cdots, x_m) = 1, \text{ for } i = 1, 2 \cdots, m$$

$$\begin{bmatrix} \frac{1-(m-1)r}{x_1}f_1 & \frac{rf_1}{x_2} & \frac{rf_1}{x_3} & \cdots & \frac{rf_1}{x_m} \\ \frac{rf_2}{x_1} & \frac{1-(m-1)r}{x_2}f_2 & \frac{rf_2}{x_3} & \cdots & \frac{rf_2}{x_m} \\ \vdots & \vdots & \vdots & \ddots & \vdots \\ \frac{rf_m}{x_1} & \frac{rf_m}{x_2} & \frac{rf_m}{x_3} & \cdots & \frac{1-(m-1)r}{x_m}f_m \end{bmatrix} \begin{bmatrix} x_1 \\ x_2 \\ x_3 \\ \vdots \\ x_m \end{bmatrix} = \begin{bmatrix} 1 \\ 1 \\ 1 \\ \vdots \\ 1 \end{bmatrix}$$

$$M\,\vec{x} = \vec{1}$$

$$M = \begin{bmatrix} \frac{1-(m-1)r}{x_1}f_1 & \frac{rf_1}{x_2} & \frac{rf_1}{x_3} & \cdots & \frac{rf_1}{x_m} \\ \frac{rf_2}{x_1} & \frac{1-(m-1)r}{x_2}f_2 & \frac{rf_2}{x_3} & \cdots & \frac{rf_2}{x_m} \\ \vdots & \vdots & \vdots & \ddots & \vdots \\ \frac{rf_m}{x_1} & \frac{rf_m}{x_2} & \frac{rf_m}{x_3} & \cdots & \frac{1-(m-1)r}{x_m}f_m \end{bmatrix} \tag{2.5}$$

$$M(\vec{x})^{-1} = \frac{1}{1-mr} \begin{bmatrix} (1-r)g_1x_1 & -rg_2x_1 & -rg_3x_1 & \cdots & -rg_mx_1 \\ -rg_1x_2 & (1-r)g_2x_2 & -rg_3x_2 & \cdots & -rg_mx_2 \\ \vdots & \vdots & \vdots & \ddots & \vdots \\ -rg_1xm2 & -rg_2x_m & -rg_3x_m & \cdots & -rg_mx_m \end{bmatrix} \tag{2.6}$$

with $g_k = \frac{1}{f_k}$.

Assignment 1

(a) Derive the equation $M\underline{X} = \underline{1}$. [Note: the right-hand side of this equation is a column matrix with all elements equal to 1. It is not the identity matrix.]

(b) Prove that the inverse of M is given by the expression shown above, assuming none of the f_i is zero.

2.2 Lee's Iteration Scheme for Solving Nonlinear Equation Set

The matrix M introduced in last chapter possesses an inverse, M^{-1}, as shown below. It is a straightforward exercise to verify directly the M^{-1} given below is indeed the inverse of M. Multiplying both sides of the matrix equation, $MX = 1$, by M^{-1} and rearranging terms in the right hand side, we can rewrite it as $X = LX$, which is referred to as the Lee Identity Matrix Equation (**LIME**), with the definition of L given below.

Definition 2.1. Define Lee Identity Matrix Equation:

$$L\vec{x} = M^{-1}M\vec{x},$$

$$\text{where } M^{-1}M = I \text{ (the identity matrix)}$$

$$L = \frac{1}{1 - mr} \begin{bmatrix} g_1 - mrg_{avg} & 0 & \cdots & 0 \\ 0 & g_2 - mrg_{avg} & \cdots & 0 \\ \vdots & \vdots & \ddots & \vdots \\ 0 & 0 & \cdots & g_m - mrg_{avg} \end{bmatrix} \quad (2.7)$$

$$g_{avg} = \frac{1}{m} \sum_{k=1}^{k=m} g_k$$

$$\textit{Fixed-point iteration formula: } \vec{x}_k^{next} = \frac{g_k - mrg_{arg}}{1 - mr} \vec{x}_k$$

This general form of the equation in X is the basis of the numerical iteration scheme for solving X. If a vector X is found such that the LHS and the RHS of the equation happen to be equal to each other, then we have found an exact solution for the set of nonlinear equations $f_i(x_1, x_2, \cdots, x_m) = 1$. If the LHS and the RHS are approximately equal to each other, then we have an approximated solution.

Consequently, the RHS of equation $X = LX$, can be used to calculate an improved estimate $\underline{X}^{(n+1)}$ from a previous approximated solution $\underline{X}^{(n)}$, where $n = 0, 1, 2, 3, \ldots$ is the order of the iteration. The process is repeated up to a large number of n and each time better solution is supposedly found. But how can we be sure?

Well, in general we are not sure. But if this process of iteration on X converges, i.e. the absolute difference between $\underline{X}^{(n)}$ and $\underline{X}^{(n+1)}$ is getting smaller and smaller as n increases indefinitely, then we have found an attractor of the algorithm, i.e. obtained a numerical solution for $X = LX$, or equivalently, the set of nonlinear equations $f_i(x_1, x_2, \cdots, x_m) = 1$.

It is the formulation of LIME that leads to the Lee's iterative algorithm for solving a given set of equations. The steps in the LIME algorithm are:

step 1 Choose the values for $\underline{X}^{(0)}$, r, and set the counter $n = 0$;
step 2 Compute $\underline{X}^{(n+1)} = L(\underline{X}^{(n)}, r)\underline{X}^{(n)}$;
step 3 Stop when $| \underline{X}^{(n)} - \underline{X}^{(n+1)}|$ is less than the pre-set tolerance;
step 4 If $n < N_0$, increase n by 1: $n \leftarrow n + 1$, otherwise stop;
step 5 Back to step 2;

where $\underline{X}^{(0)}$ is the initial guess, r is a search parameter and N_0 is the maximum number of iterations.

If this procedure stops at step 3, we have $|L(\underline{X}^{(n)}, r)\text{-}I | \sim 0$, i.e. $L(\underline{X}^{(n)}, r)$ becomes very close to the identity matrix, where N is the value which n equals to when the exit condition in step 3 is satisfied. If this procedure stops at step 4, then the procedure does not converge within N_0 steps and we have to either increase N_0 or choose another $\underline{X}^{(0)}$ and restart the iteration again.

A few observations can be made at this point to characterize the LIME algorithm:

(1) The algorithm is an iterative scheme with 2 input parameters to start the process, the initial guess $\underline{X}^{(0)}$ and r.

(2) The algorithm relies only on the evaluation of $f_i(x_1, x_2, \cdots, x_m)$ and not on any the their derivatives.

(3) Convergence of the algorithm depends on the choice of $\underline{X}^{(0)}$ and r.

(4) The set of unknown variables $(x_1, x_2, \cdots, x_m)$ under consideration has to be unbounded since as the iteration scheme continues $(x_1, x_2, \cdots, x_m)$ can be anywhere in the entire m-dimensional space.

$$n = iteration$$

$$x_k^{(n+1)} = \frac{g_k - mrg_0}{1 - mr} x_k^{(n)}$$

$$g_k = \frac{1}{f_k(\bar{x}^{(n)})}$$

$$g_0 = \frac{1}{m} \sum_{j=1}^{m} g_j(\vec{x}^{(n)}).$$

Assignment 2

(a) Prove the Lee Identity Matrix Equation: $\underline{X} = L\underline{X}$, with L defined above.

(b) Prove that if the iteration scheme shown above converges, i.e. Lim $|\underline{X}^{(n)} - L(\underline{X}^{(n+1)}| \to 0$ as $n \to \infty$, then we have found a solution of the original set of nonlinear equations: $f_i(x_1, x_2, \cdots, x_m) = 1$.

(c) Convince yourself by deriving an expression how the rate of convergence is controlled by the parameter 'r'.

2.3 Fine-Tuning the Path to Convergence

The rate of convergence of the iterative scheme introduced in the last chapter depends strongly on the parameter 'r'. Therefore, we refer 'r' as the rate parameter. However, just having one searching parameter is not enough, especially for large m, since once we choose a starting point $X^{(0)}$ and 'r' the path to the successive $X^{(n)}$ is fixed. Effectively, an m-dimensional search for the optimal point is collapsed to a 1-dimensional search. This has the advantage of simplifying the search process on the one hand, but on the other hand, it limits the flexibility to locate the global optimal point in a reasonable time. For this reason, we proceed to introduce another searching parameter to enrich the freedom or flexibility of the searching process.

In many applications, we have some ideas where the solution of our problem roughly located due to previous experience or an educated guess. In such situations, we are not searching for the root blindly and aimlessly, wandering all over the entire space. Instead, we are searching in a finite domain and the set of equations to be solved is: $f_i(x_1, x_2, \cdots, x_m) = 1$, for $i = 1, 2, 3, \cdots, m$, with b_k are variables bounded between the fixed boundaries given by b_{k+} and b_{k-}, i.e. $b_{k-} < b_k < b_{k+}$. Such a set of bounded variables, $(b_1, b_2, \cdots, b_m)$, can be transformed to another set of unbound variables $(a_1, a_2, \cdots, a_m)$ by the following:

$$a_k - a_{k0} = tan\left(\frac{\pi(b_k - b_{k0})}{b_{k+} - b_{k-}}\right)$$

where $b_{k0} = \frac{(b_{k+} + b_{k-})}{2}$ and a_{k0} are constant offsets for $k = 1, 2, 3, \cdots, m$.

This transformation maps the bounded domain (b_{k+}, b_{k-}) to an unbound domain. By substituting the inverse transformation into the set of given equations it becomes: $f_i(x_1, x_2, \cdots, x_m) = 1$, for $i = 1, 2, 3, \cdots, m$, which can be solved by the iteration scheme presented in the last section.

The net effect of the consideration above is to introduce a path parameter 'p' into the search scheme, as shown below, in addition to the 'r' we already have. So an m-dimensional search is now collapsed to a 2-dimension search in (r, p).

> **Definition 2.2.** Since $\Delta x_k^{next} = \frac{x_k^{next} - x_k}{x_k} = \frac{(g_k - 1) - mr(g_{avg} - 1)}{1 - mr}$,
>
> $$\Delta x_k^{next} = (g_{avg} - 1)$$
>
> When $g_{avg} \to 1$, $\vec{x}$ converges.
> Define r and p as the root and path convergence control parameter with $g_k(\vec{x}, p) = \frac{1+p}{f_k + p}$ of a fixed-point iteration.ples.

Assignment 3

Show that the parameter 'p' introduced in this section do not affect the convergence of the iterative scheme. However, it will affect the path in the m-dimensional space that starts from the initial guess to the point which the iterative process converges to.

2.4   From Equation Solver to Optimizer

The iterative algorithm derived from the LIME to solve a set of nonlinear equations $f_i(x_1, x_2, \cdots, x_m) = 1$ has an immediate application to find the local maximum or minimum of any continuous differentiable function of n variables.

> **Definition 2.3.**
> Given a continuous differentiable function $F_{obj}(x_1, x_2, \cdots, x_m)$, we can define $f_k(X) = (\frac{\partial F_{obj}}{\partial x_k})^2 + 1$ for $k = 1, \cdots, m$; where the upper case X is used as a shorthand for the vector $(x_1, x_2, \cdots, x_m)$.

By finding the roots of nonlinear equations $f_k(X)$ we obtain the local maxima or minima of F_{obj}, the objective function of interest. The figure below illustrated the entire iteration process.

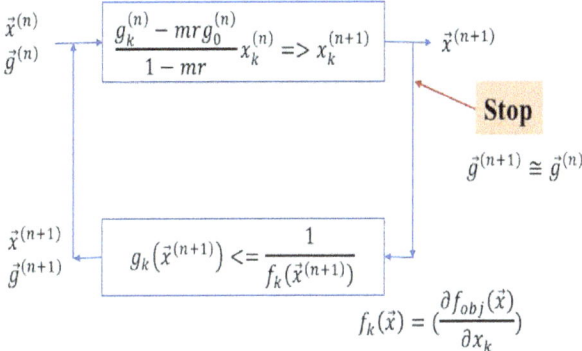

Fig. 2.1:   Iteration process.

Summary of Chapter 2

The key points that can be gleaned from this chapter are

- A system of nonlinear equations with m unknowns can be transformed into a matrix equation, which serves as the basis of an iterative scheme to generate successive approximations of the solution of the nonlinear equation system. This iterative scheme, if it is convergent, is a mapping moving a point within the Basin-of-Attraction eventually to a solution of the nonlinear equation system.
- An adjustable parameter 'r' is introduced in the matrix equation, hence also in the iterative scheme, to control the rate of convergence of the scheme.
- An additional parameter 'p' is further introduced to control the path in the m-dimensional space to search for the answer that the iterative process finally converges to.
- This procedure can be applied to a bounded m-dimensional region (a constrained search) where we have some prior knowledge about where the solution should roughly be.
- Given an initial guess within the constrained space around the solution, an iterative search can be performed to find the exact location of the solution by choosing a pair of (r, p) parameters. In the non-adaptive version, (r, p) would be fixed during the iteration, and the objective function would be a function in m-dimensional space.

- In the general case, the (r, p) pair is regarded as additional variables in the objective function so that the objective function would be a function in the $(m + 2)$-dimensional space. Hence, this procedure is named ACORN, an acronym from "Adaptive Constrained Optimal Robust Nonlinear". It is 'Adaptive' because the values of r and p are modified during the iterative process, and 'Constrained' because the procedure is applied to a bounded m-dimensional region.

Chapter 3

Application to Pedagogical Examples

In this chapter we will discuss how the ACORN optimization algorithm distinguishes itself from the rest of the pack available commercially by simple examples.

In the following example, we demonstrate the efficiency and robustness of the ACORN algorithm by fitting a set of data to a function of the following form

$$F(a, b, y) = a * cos(b * y) + b * sin(a * y), \qquad (3.1)$$

where a and b are parameters to be determined by the fitting process and y is a 1D independent variable which we sample. This pedagogical example is used to test the classic optimization algorithm due to Levenberg and Marquardt[1].

We pick it because it provides a stark contrast to show the superiority of ACORN compared with the LM algorithm and others with similar philosophy. The codes implemented for this example is listed in the Appendix A, in Octave. Equation (3.1) is chosen due to the sensitivity of the final answer obtained from the optimization process to the initial guess of parameters 'a' and 'b'. For this reason, it provides a good test to the efficiency and robustness of ACORN, or any other optimization algorithms of similar kind, in spite of the fact that it is pedagogical in nature.

The driver code in this example is `fitSinCos.m` in Appendix A, which calls `Aimigo8.m` to drive the ACORN package. In the discussion below, the mathematical problem posed in front of a data modeler is the following.

Suppose all we know is a sample data of y_i and their corresponding $F(y_i)$, where F is the observed dependent variables. We suspect that the relationship between y_i and $F(y_i)$ is of the form given by Eq. (3.1), but the

---

[1]It can be found in Wikipedia.

parametric pair (a, b) is unknown to us. The obvious problem posed to any data modeling practitioner is: Can we determine the correct pair of (a, b) from the knowledge of the dataset y_i and $F(y_i)$ we have sampled?

The answer is a definite 'yes', and the methodology is mathematically known as *regression analysis* [Myers (1990)].

Since the parameter pair (a, b) is unknown to us, it is a classic case of *least-square* fit. The objective function $G(a, b)$ is defined as the "sum over all j of the squared errors" (SSE):

$$G(a, b) = SSE = \sum_{j} [F(y_j) - F(a, b, y_j)] * *2. \qquad (3.2)$$

As a concrete example, we first set the parameter pair (a, b) in Eq. (3.1) arbitrarily equals to $(17.9, 16.9)$. This set of values is not as extreme as in the example used in the *Wikipedia* page, where 'a' and 'b' are set to values over 100. However, it is enough to show the difficulty facing by the LM type algorithms, without straining the graphical clarity in the figures below.

If there is no noise in the measurements of $F(y_j)$, the minimum of $G(a, b)$ will be exactly zero and occurs at $(17.9, 16.9)$. Hence by minimizing $G(a, b)$, in principle we can determine the correct value of the (a, b) pair.

However, as shown in the following sequence of 3-D plots the actual objective function $G(a, b)$ depends also on the actual data sample we are using. Different set of sample data will lead to a different objective function. But of course, the supposedly unknown set (a, b) is always the same [in our case it is $(17.9, 16.9)$]. The question is: when we try to determine (a, b) from these different sample data (hence different objective functions) using *least-square* fit, will we recover the *same* and *correct* set of (a, b)?

The sequence of plots in Figs. 3.1 show that for the same problem the objective functions can be vastly different, depending on the sample data we analyzed. In each of the 3-D plots below, there is ONE global minimum (which is zero) that correctly gives us the right answer $(17.9, 16.9)$. However, the number of local minima and their locations are all (data) sample dependent. If the optimization algorithm we use is unable to locate the global minimum, then we end up with one of the local minima depending on the data sample under study and totally miss the *correct* answer. These local minima are spurious and entirely depend on the data we can get our hands on. From the trend shown in Figs. 3.1(a) to 3.1(c), we come to the conclusion that the more data we are using in the fitting, the more spurious local minima we are getting, making it more difficult to converge to and easier to miss the global minimum.

Unfortunately, with most optimization algorithms we have no idea whether the minima they converge to are global or local. An example we have encountered is the `fmincon` routine in MATLAB.

In Figs. 3.2(a) and (b), the basins of attraction for the objective function in Fig. 3.1(a) are displayed in different colors. In Fig. 3.2(a) the BOAs of the "interior point" algorithm in `fmincon` are shown, while in (b) the BOAs of the "sqp" algorithm are shown. In both (a) and (b) we show not only the BOA of the global minimum (in red), but also those for two other major local minima on the LHS and RHS of the global minimum (in blue and green). As shown in Fig. 3.1(a), there are actually many local minima in this objective function so we cannot show the BOAs for all of them. The white region is the set of all initial guesses that do not converge to the global minimum and the two other major local minima. These two BOA plots look not very different on the whole and are more complicated than those shown in Figs. 1.2(a) and (b), because we have a lot more local minima in this example. There is some evidence that the basin boundary is a *fractal* set, i.e. the structures of the basin boundary are very intricate, and fine-scaled. The plots we produced here is not fine-scaled enough to reveal all the intricacies. This may explains why the initial guess has to be very close to the global minimum in order for the two algorithms in fmincon to converge to the global minimum. Otherwise, the algorithms converge to one of the local minima and return the wrong answer. This is an example of very sensitive initial conditions for the Levenberg-Marquardt and related algorithms, such as "interior point" and "sqp".

We will not further burden our reader to any graphic display of the basins of attraction for the objective functions in Fig. 3.1(b) and (c), since it will be guaranteed to be more complicated and confusing. Figures 3.2 have already sent a strong message that how difficult it will be to come up with a correct *least-square* fit if we do not have a good optimization algorithm that guarantees us the global minimum.

In the example above, our objective function is defined in 2-D space, so that there is no difficulty for us to show and visualize the BOA graphically. However, for an objective function defined in n-dimensional space, it is impossible for us to achieve the same effect. We may wonder if we are going to lose all our insight of the optimization process.

Nonetheless, the BOA concept is a useful tool to delineate the working of an optimization algorithm, specifically, how the input affects the output of the algorithm. However, it is extremely computationally intensive to generate a BOA plot for any practical application, even in 2-D.

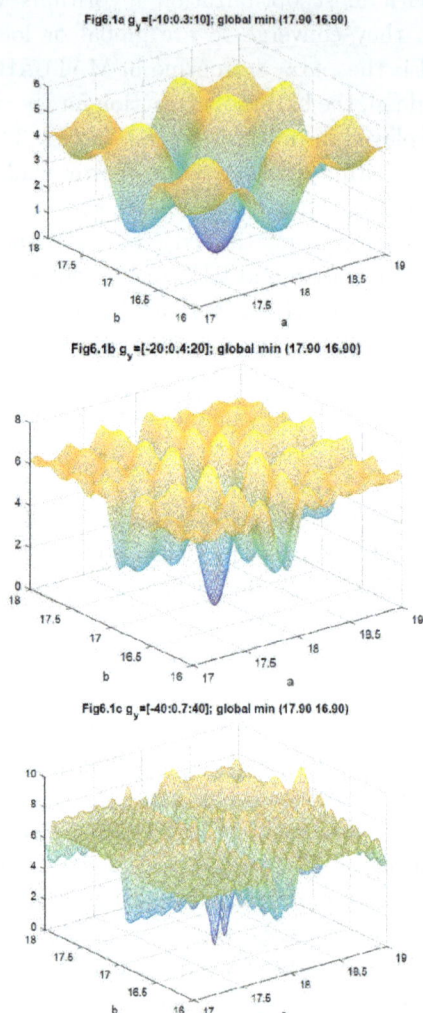

Fig. 3.1:   The 3-D plots of the objective functions all with global minimum located at $(a, b) = (17.9, 16.9)$, are the sums square errors of sample data points obtained from vector of data y_j spanned from (a) $[-10 : 0.3 : 10]$, (b) $[-20 : 0.4 : 20]$, and (c) $[-40 : 0.7 : 40]$, where $[e_1 : d : e_2]$ stands for a vector with first element equals e_1, last element equals or less than e_2, and successive elements differ by d. These objective functions all have same global minimum equal to zero located at $(17.9, 16.9)$. But they have very different (non-zero) local minima located at vastly different places, and the more data we used in the fitting, the more spurious local minima we got. Obviously, if the optimization algorithm does not converge to the global minimum, then the least-squares fit will give the wrong answer.

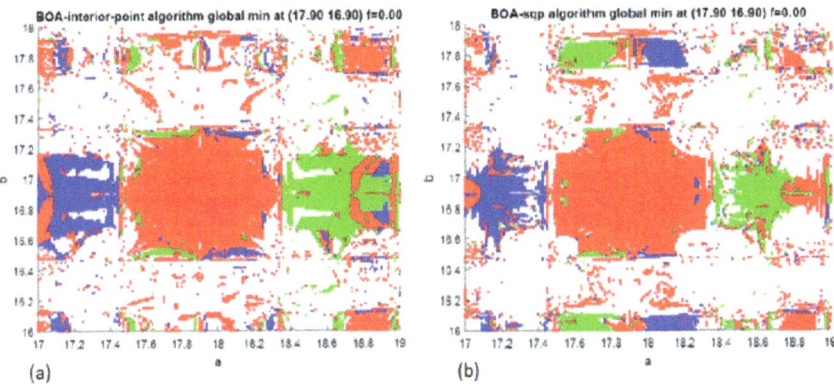

Fig. 3.2: The basins of attraction for the objective function in case of Fig. 3.1(a), where the objective function looks simpler. In (a) the BOA of the "interior point" algorithm is shown, while in (b) the BOA of the "sqp" algorithm is shown. In both (a) and (b) we show only the BOA of the global minimum (in red), while those for two other major local minima on the LHS and RHS of the global minimum are also shown (in blue and green). As shown in Fig. 3.1(a), there are still many local minima in this objective function even it is simpler than Fig. 3.1(b) and (c), so we cannot show the BOA for all of them. The white region is the set of all initial guesses that do not converge to the global minimum and the two other major local minima. These two BOA plots do not look very different on the whole and are more complicated than those shown in Figs. 1.2(a) and (b), because we have a lot more local minima in this example. There is some evidence that the basin boundary is a fractal set, i.e. the structures of the basin boundary are very intricate, and fine-scaled. This explains why the initial guess has to be very close to the global minimum in order to converge to the global minimum. Otherwise, the algorithms converge to one of the local minima and return the wrong answer.

Fortunately, the special way of how the ACORN algorithm works really helps us to overcome this difficulty. In contrast, BOA of the ACORN algorithm is an extension of the original concept to its 2-D projection. In the ACORN algorithm, we search in the 2-D parameter space (r, p), where 'r' controls the convergent rate and 'p' controls the convergent path, regardless of the dimensionality of the problem. The BOA in ACORN is defined in the 2-D (r, p) space, instead of the n-dimensional space of the objective function where we are looking for the global minimum as in the previous example, and it provides us in principle with the same information and insight on all the local and global minima as the BOA in n-dimensional space. In particular, the global minimum can be easily identified in such 2-D (r, p) search. Hence, the chance for ACORN algorithm to find the global minimum is higher than any of its competitors. However, in practice due to the finite size of the (r, p) grid and range of search, we may miss the global minimum occasionally.

The BOA in ACORN is determined as described below. For a given initial vector, X, in the n-dimensional space as the starting point of the search for global minimum, we perform a scan in the 2-D (r, p) space. The choice of the initial point in (r, p) space to start the ACORN algorithm eventually determines the attractor of the procedure, regardless of the value of n.

For simplicity in argument, we illustrate the concept of BOA with a non-adaptive version of ACORN in which the (r, p) pair is not evolved adaptively during the iterative process, even this is not used in practice due to its limitation. Each (r, p) pair is eventually mapped to an attractor in the n-dim space by the ACORN algorithm and the (r, p) pair is not changed in the process. The systematic 2-D scanning of the initial (r, p) space will eventually define the "BOA" of the ACORN algorithm, an example of which is shown as a 3-D plot in Fig. 3.3, where the horizontal plane is the 2-D (r, p) space and the vertical axis is the value (in log 10 scale) of the objective function evaluated at the attractor corresponding to each of the initial (r, p) pair. Figure 3.3 is borrowed from the concrete example discussed below to illustrate how the concept of BOA in the ACORN algorithm is different from what we have encountered so far. In fact, we can regard it as an unconventional kind of BOA or an extension of the normal BOA concept, whose sole purpose is to contrast with the conventional BOAs for traditional optimization algorithms.

In an adaptive ACORN algorithm, which is used in all our discussions in this work, the picture of BOA is not that much different from what is shown in Fig. 3.3, even the initial (r, p) pair will evolve along with the

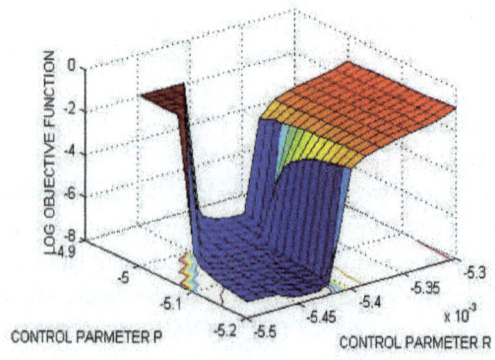

Fig. 3.3:   A typical BOA in (r, p) space.

n-dimensional X vector as the iteration moves ahead. All we need to keep in mind is that the horizontal (r, p) plane in Fig. 3.3 is the space of the initial (r, p) pairs to start the iterative process. Eventually when the process converges to a solution the final (r, p) pair will be different from its starting value. The net effect remains the same as in a non-adaptive ACORN algorithm, i.e. each initial or starting (r, p) pair is eventually mapped to an attractor in the n-dim space by the ACORN algorithm. Each attractor is a point in the n-dim space that corresponds to a minimum of the objective function uncovered by ACORN. By examining the surface in the 3-D plot similar to that as shown in Fig. 3.3, we have a much clearer picture of where the global minimum is *potentially* located. We judiciously used the words *"potentially located"* in the previous sentence to hedge our statement because there is one more factor that can affect the picture of BOA that we have not considered so far. The BOA plot in our discussion up to this point is based on a given initial guess in the n-dim space of X. Given another initial guess of X, we will get a different BOA plot, which may result in lower or higher minima. Fortunately, in practice, the BOA plots never depend sensitively on the initial guess of X, if we already have some idea on the location where the minimum is. This experience really comes as a godsend, and effectively it collapses the n-dimensional search for global minimum to a 2-D search in the (r, p) space.

We repeat below the solution to the least-square fit problem with model equation given by Eq. (3.1), using the ACORN algorithm. In this case, the correct answer in (a, b) are deliberately chosen to be much higher values than the one used in Figs. 3.1 and 3.2, $(101.13, 102.04)$, and the data sample is generated from -40 to 40 with a uniform separation of 1, so it is a much more difficult problem than what we have discussed in the first part of this chapter, since the local minima are cramped together very densely. The 3-D plot of the objective function in this case will be very similar to Fig. 3.1(c), except that there are more minima and maxima cramped around the global minimum located at $(101.13, 102.04)$.

In Fig. 3.4, we show how the objective function (in log 10 scale) converges as ACORN iterates in a sequence of scans in 'r' with a fixed $p = 0$. For example, when the initial r is set at -0.001 (the red line) eventually the objective function is reduced from 1 to below 0.01, after 25 iterations. This game is repeated for other values of 'r' systematically. Out of this maze of 'r' scan, we discover that when the initial 'r' is set at 0.89 (green line), initially it looks like this is a bad choice because the objective function increases by almost a 100 fold before it sinks to about $\frac{1}{10}$ of its initial value.

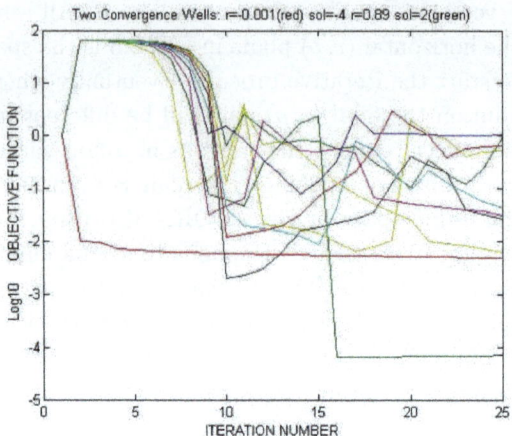

Fig. 3.4: A systematic scan in (r, p) space started from an initial guess. The green line, representing the scan with initial $p = 0$ and $r = 0.89$ and produced the deepest reduction in the objective function, is most likely where the global minimum located.

But ultimately, it achieves an amazing reduction of the objective function from 1 to below 0.0001, which means that the initial setting of $p = 0$ and $r = 0.89$ converges to a solution inside the BOA of the global minimum. On the other hand, the pair $p = 0$ and $r = -0.001$ converges to a solution within the BOA of a local minimum.

In this example, the systematic 2-D scanning of the initial (r, p) space eventually led us to uncover the location of global minimum.

Figure 3.5 shows the 2-D contour plot of the objective function in this example, where the correct answer in (a, b) are of much higher values than the one used in Figs. 3.1 and 3.2. The contours in deep blue to black represented the lower end in the values of the objective function, and those in red to dark red corresponded to the peak values of the range. The global minimum is clearly shown near the right-center of the figure. There are also a number of local minima distributed along a vertical and a horizontal band, just as what we can visualize from Figs. 3.1 earlier. Using the best pair of (r, p) obtained in Fig. 3.4 and an initial guess located at the small red circle near the center of the figure, the ACORN iteration was initially influenced by the local minimum and moved towards the local minimum indicated by the blue circle. It looks like it will end up being trapped by the local minimum. However, due to its systematic scanning of the (r, p) space, ACORN somehow could look ahead and decided there was a better chance

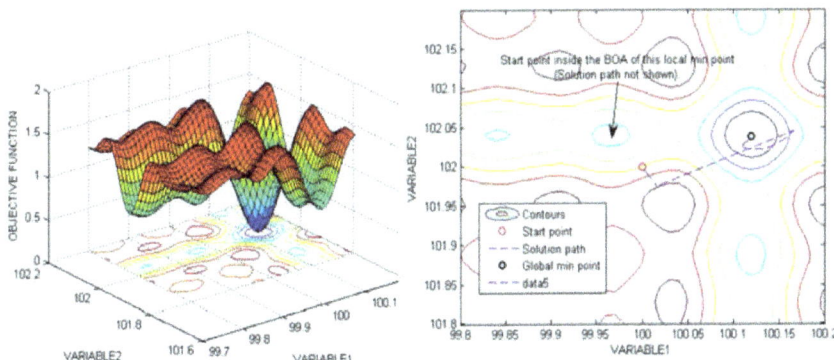

Fig. 3.5: 2-D and 3-D Contour plots of the objective function with the global minimum at $a = 101.13$, $b = 102.04$ (VARIABLE1 is a, and VARIABLE2 is b, as shown in the plots), showing peaks (in red) and valleys (in blue and black) in the objective function. Starting from an initial guess (red circle near center of plot), the ACORN algorithm performs a sequence of (r, p) scan to look for global minimum. This example showed how ACORN jumped out from a local minimum and eventually located the global minimum.

of finding the global minimum somewhere else. Hence, it made a surprising move, jumped out from the shallow well and moved to the vicinity of the global minimum. From that point onward, it took just two more steps for the algorithm to lodge firmly in the global minimum.

In the next graph, Fig. 3.6, where the 3-D plot of the objective function with global minimum located at $a = 100.8$ and $b = 100.4$, as defined in Eq. (3.1), is shown. The presence of large number of local maxima and minima in this plot is due to the large dataset we used to construct the objective function. As a rule of thumb, the number of local maxima and minima is roughly proportional to the number of data involved in the problem.

We started the ACORN search at the center, $a = 100$ and $b = 100$. In case the readers may wonder why it is such a big deal to show this search again, we take the pain to point out that the initial guess where ACORN started the search is many local minima away from the global minimum, even just by numerical values it does not appear to be that far away. For gradient descend type of optimization algorithms, such an initial guess will definitely end up hopelessly trapped in one of the many local minima.

After the search in the (r, p) space is repeated 8 times, the objective function was reduced by nine order of magnitudes and the global minimum was found by ACORN without a doubt. The convergence plot of the objective function is shown in Fig. 3.7.

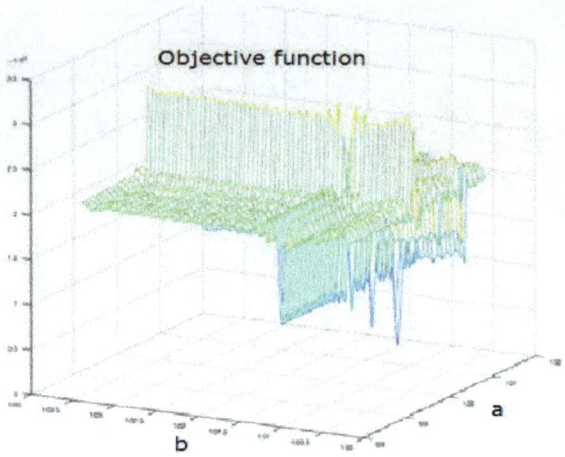

Fig. 3.6: 3-D plot of the objective function with global minimum located at $a = 100.8$ and $b = 100.4$ [Eq. (3.1)], with ACORN search started at $a = 100$ and $b = 102$.

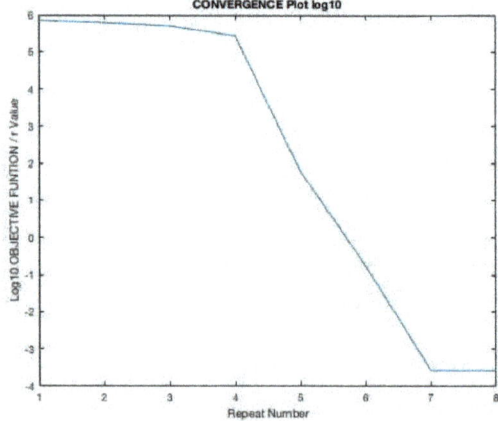

Fig. 3.7: Plot of the lowest values of objective function (shown in Fig. 3.6) achieved by repeating the ACORN (r, p) search 8 times.

To summarize, the examples we have discussed in this chapter is deliberately chosen to be simple and easily visualized. They are optimization problems with 2-D objective functions, so that we have the luxury to inspect the behavior of the objective function in detail. Yet, the surprise that emerges from our investigation is that local minima in an objective function of a *least-square* fit problem is data dependent and hence can be regarded as spurious. Only the global minimum is the correct answer to the *least-square* fit problem. We do not claim that this phenomenon is universal since we do not have a rigorous mathematical proof. However, our examples here definitely serve as an existential proof. If this kind of phenomenon can occur in a simple 2-D objective function, we can be sure that it would be more likely to occur in higher dimensional landscape where we lost our ability to visualize the detail behavior of the objective function and complexity of the problem can be much higher.

In the next chapter, we will discuss this phenomenon in a 6-D space example.

Summary of Chapter 3

The key points that can be gleaned from this chapter are

- Using a simple example in regression analysis, we can demonstrate that the detail behavior of an objective function with just two unknowns depends on the dataset being used. The number of local (fictitious) minima and their locations depend strongly on the dataset. Only the global minimum is the true solution to the problem. Also demonstrated is the BOA of the global minimum for LM type algorithms is very close to the global minimum itself, hence it is easy for these types of algorithms to get trapped in one of the local minima without knowing it.

- For an objective function defined in n-dimensional space, the BOA of the ACORN algorithm is defined in the (r, p) space, which can be regarded as a projection of the n-dimensional BOA to the (r, p) space. The 2-D project of BOA into (r, p) space will give us a clear picture of where the global minimum is.

- For an initial guess X_0 in the n-dimensional space, the objective function can be optimized by scanning a 2-D grid in (r, p) space. The minimum found each time can be used as a new guess to repeat the scanning procedure, until the required accuracy has been reached. This provides a systematic way for the ACORN algorithm to track down the global minimum.

Chapter 4

A Data-Modeling Example from Accelerator Physics

In this and later chapters, we will discuss examples of practical applications of the ACORN algorithm.

In high energy physics experiments, measuring the high energy particle beam profile inside the accelerator beam-line is the first step to quantify the beam quality achieved by the accelerator. The beam profile is measured by sticking metallic wire inside the vacuum chamber of the beam-line to obtain 1-D or 2-D beam intensity profiles. From the raw data, physicists attempt to build mathematical model to describe the beam intensity profile so that they have a better description and characterization of the beam.

In the following, we look at a set of one dimensional raw data from such a measurement of beam intensity, provided to us by physicists at the Institute of High Energy Physics (IHEP), Chinese Academy of Science (CAS).

A natural mathematical fit to such profiles is a one-dimensional Gaussian function, characterized by three parameters: the location of the peak, the height of the peak, and the half-width (or the standard deviation). However, particle beams are often split and recombined during the accelerating process. Most likely the beam profile cannot be described by a single Gaussian, and at least a double Gaussian should be used.

In a classical *least-square* fit of the raw data to a sum of two Gaussians, the six parameters have to be determined by minimizing the sum squared-errors (SSE), the objective function in this case.

Figure 4.1 shows a double Gaussian fit (blue line) to the raw data (red dots) by using the `fmincon` routine in MATLAB to minimize the objective function (SSE) with the default "interior point" algorithm. The parameters for this double Gaussian fit is `b=[7.4718400 1343.3837648 ...` `-4.3877400 0.3480186 2085.5936249 4.5956212]`, which stands for the background level, amplitude, location and half-width of the first

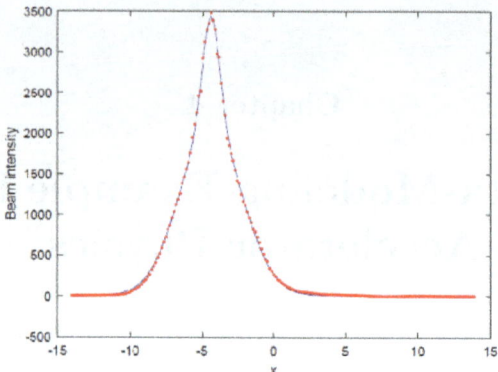

Fig. 4.1: Raw data (red) and numerical double Gaussian fit (blue), using MATLAB routine `fmincon`, of a high energy particle beam profile, with parameter b = [7.4718400 1343.3837648 −4.3877400 0.3480186 ... 2085.5936249 4.5956212] as defined in the text. Beam data and fit are provided by physicists in IHEP of CAS.

Gaussian, followed by the amplitude and half-width of the second Gaussian. In this particular fitting model, both Gaussians are assumed to be centered at the same location. Incidentally, it happened that the physicists in IHEP-CAS used the optimization routine in *Mathematica* and got the same result. *Mathematica* is a favorite software tool of high energy physics community because its founder and chief developer, Stephen Wolfram, started his career as a promising young high energy physicist.

In either case, we have no idea the answer we got from MATLAB or *Mathematica* is the global minimum or a local minimum, since neither algorithm is able to give us that information. And as we have seen in the last chapter, in this kind of fitting problem there can be many fictitious local minima introduced by the particular data sample we used. There is only one global minimum that gives us the correct answer.

To satisfy our curiosity, we would like to investigate if there are more than one minimum can be found in this least-square fit problem by ACORN. Unlike the 2-D problem we dealt with in the last chapter, here we have an objective function in 6-D space. So we are unable to visualize the behavior of our objective function graphically in this 6-D space. We deliberately chose an initial guess, so far away from the BOA of the MATLAB *(Mathematica)* solution that MATLAB *(Mathematica)* failed to find the best-fit solution, to start the ACORN optimization algorithm.

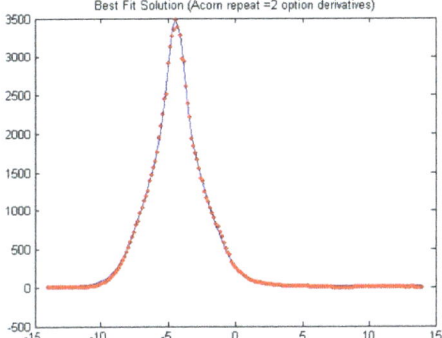

Fig. 4.2: Raw data (red) and a different numerical double Gaussian fit (blue) with parameters b = [12.6004864 1247.9375313 −4.3860236 0.3124490 ... 2183.4826927 4.2760951], using ACORN, of a high energy particle beam profile. The difference between this fit and the one shown in Fig. 4.1 is barely observable, but it is still visibly perceptible by carefully inspecting the fits near the peak of the profile.

With our initial guess, ACORN was able to find a slightly different fit as shown in Fig. 4.2 with the parameters b=[12.6004864 1247.9375313 ... −4.3860236 0.3124490 2183.4826927 4.2760951], obviously a solution different from the one shown in Fig. 4.1. Even the difference between the two fits is very small, it is still visibly perceptible by carefully inspecting the fits near the peak of the profile. This ACORN solution seems to fit the peak better.

Next, we ran ACORN again to see if it can find more than one best-fit solutions. After repeating the running of ACORN many times with a different set of (r, p) values as the start points, it was able to find a second best-fit solution with a slightly higher objective function than the one found by MATLAB (Mathematica), and parameters b=[7.6396166 1336.5131930 ... 4.3877104 0.3436117 2094.0375221 4.5709423], which is not exactly identical to the MATLAB *(Mathematica)* solution but close enough to be regarded as the same solution, after taking into account the difference in algorithms and machine round-off errors. The resulting fit is almost identical to that shown in Fig. 4.1.

It is a bewildering experience to find more than one fit that is about the same and seems to be equally acceptable. Practically, a cavalier data modeler would be happy to find either and stop looking for more answers. We may attribute this to the resilience of the physical problem at hand because the global and local minima found by different algorithms all produce good enough fits that are roughly the same.

Mathematically, this is understandable based on what we have found in the examples shown in the last Chapter. As the number of data used to define the objective function becomes larger, the number of local minima clustered around the global minimum increases also. Hence, even an optimization scheme does not necessarily returns the global minimum, the answer it produced, one of the local minima that is close to the global minimum, is often a good approximation of the global minimum.

However, on a deeper level, the existence of more than one solution to an optimization problem is unsettling. We have no proof, but we suspect that the situation is similar to the 2-D examples we studied in the last chapter. The 6-D objective function we tried to optimize in this case is a lot more complicated than we can imagine, with many local minima densely distributed very close to the global minimum. Furthermore, the global minimum is just a slightly deeper well in that perplexing landscape. We are lucky in this case that the two solutions we found gives us apparently equally good fits. But there is no guarantee that we will be lucky again in another application.

Which one is a better fit? Most accelerator physicists would prefer a fit that gives a better model around the peak of the beam profile, since that is where the action is. In that case, the baffling situation is a blessing in disguise, because it serves as a warning or reminder for us to rethink what our goal is. If a more accurate fit around the peak of the profile is important to us, then the objective function should not be a simple sum of all squared errors with equal weighting. The squared errors around the peak should be given higher weighting factors compared with those errors from the tails.

At this point, we have to stop our analysis on this problem here because a more rigorous and complete solution to this beam profile modeling problem should include other mundane considerations such as the error-bars associated with the measurements, which will also affect the weightings in the sum squared errors (objective function).

The discussion in this chapter reveals the true value of the ACORN optimization scheme compared to its peers. Most canned optimization routines available commercially are black-boxes that accept your objective function together with the constraints you specified. After all the dust is settled, it will either split out no answer at all or an answer that you have no idea that it is a local or the global minimum.

As it is well known to many experienced practitioners in data modeling, there are no "closed form" solutions in nonlinear models. There is the ever-present danger of finding locally optimal rather than globally optimal solutions for problems under investigation.

In contrast, ACORN optimization scheme will most likely give its user the global minimum rather than other local minima as a result of its 2-D (r, p) search approach. We gain more insight to the nature of the objective function in the domain of interest. Sometimes, this knowledge may lead us to rethink the problem and revise the objective function to one that serves our goal better. This adds a new dimension to the process of data-mining and machine learning, in which human (the researcher) are now an integral part of the man-machine loop that iterates to deeper and deeper learning and understanding. The deeper knowledge gained in this process will likely to give us an edge to mine for the real gems in the data we are studying.

Summary of Chapter 4

The lesson that can be learned from this chapter is

- Through a specific particle beam profile fitting problem, the results from ACORN are compared to those obtained by MATLAB and Mathematica. In the example discussed, ACORN is able to find the global minimum that gives a better fit, as well as another good but not better fit that is also found by MATLAB and Mathematica. The global minimum, which corresponds to the best fit, is found by ACORN but not by MATLAB, nor Mathematica. This is a clear case that the optimization algorithms used in MATLAB and Mathematica are trapped by local fictitious minima and not able to get out.

Chapter 5

Applications to Machine Learning and Neural Networks

5.1 Artificial Neural Networks — A Brief Review

Before we touch upon the topic of how ACORN can contribute to the improvement of Artificial Neural Network technology, we review briefly what exactly it is. However, for serious readers, the following is definitely not a replacement of more complete treatment in the subject[1].

An Artificial Neural Network (ANN), as its name implies, is a mathematical model that relates a set of input signals (data) to another set of output signals (data) based on our understanding of how a biological brain responds to external stimuli from sensory inputs. Just as a brain evolves into network of interconnected cells (neurons) to form a massive parallel structure to efficiently process information it receives, ANN organizes a network of artificial neurons (nodes) into a similar structure to solve the learning problem. Input data are fed into a group of receiving artificial neurons known as the input layer, which connects to one or more hidden layers in between, before finally to the output layer that communicates the network's output, to its user, based on which proper response can be taken.

The connections between neurons in each layer are known as synapses. In its simplest form, information in an ANN flows straightly forward from the input and ultimately to the output layer. This type of ANN architecture is known as a Feedforward Neural Network. It mimics the way how the brain processes input sensory data into recognition or decision as output. Nodes within the same layer are not connected, and they process the data received independent of each other, extracting the specific features they need from the incoming data. However, each node is connected to all the nodes in the previous layer, from which they receive the signals, and forward the

---

[1]Such as [Haykin (2011)].

processed information to all the nodes in the layer ahead. Each connection between the nodes has a weight associated with it. During the learning process, these weights are constantly adjusted as data flowing through the nodes, so that the final output matches some presumed answers as close as possible.

The exact number of nodes within a layer and the number of hidden layers in between the input and output layers are problem dependent, i.e. they are different in different applications. It is a judgement call by the users according to their experiences and gut-feelings, or following examples of similar applications. In most cases, users of ANN have to fiddle around to come up with the proper architecture for the problem they try to solve. An example of the feedforward type of ANN with just one hidden layer, which is historically also known as a perceptron, first proposed by Frank Rosenblatt in 1957, is shown schematically in Fig. 5.1.

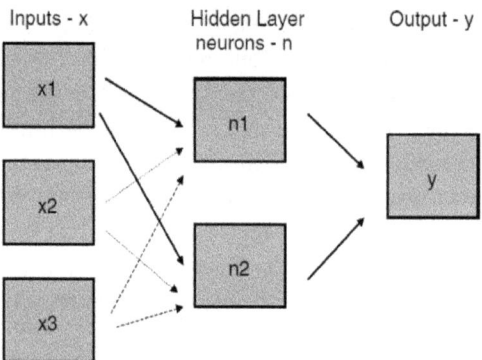

Fig. 5.1: A typical feedforward neural network with 3 nodes in the input layer, 2 neurons in the hidden layer and 1 node in the output layer.

The difference in the final synaptic weightings (at the end of the learning process) reflects each node's preference of valuing some stimuli over others. Take the example of a child learning how to recognize an apple. An apple is round in shape and it can be red, green or yellow in color. It has a certain size, similar to an orange or a pear. But more importantly it has a smooth and crispy skin and a distinct taste (and texture), even sometimes it can be a bit sour instead of sweet. All these features distinguish it from an orange, pear or peach. Thus, the weights for these different stimuli to the nodes in the hidden layer are gradually changed as the child is exposed to more and more examples of apples, strengthening some synaptic connections while weakening others until eventually the child can accurately identify an apple.

An artificial neuron processes the input data first by forming linear combinations of the input data according to the weights associated with the synaptic connections, then by passing these linear combinations through the activation (or transfer) function to produce the output from this node, and send it to the nodes in the following layer, hence introducing nonlinearity into the ANN. The activation function is acting like a switch to let information passing through the node if the linear combination of all the incoming stimuli is larger than a certain threshold. Different forms of activation function have been proposed by different authors. Popular examples of activation function include the logistic function, the hyperbolic tangent, and the arctangent functions, because of their smoothness and differentiability properties. They all share similar characteristics that allow them to mimic the firing and non-firing states of a biological neuron in our brain or the "on-off" switch in a digital network. This manipulation of incoming data by a single node is illustrated schematically in Fig. 5.2.

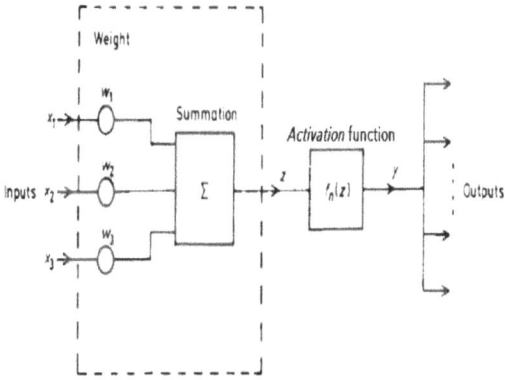

Fig. 5.2: Input and output data of a single node.

Effectively, each node in the hidden layer extracts a specific feature from the input raw data by its own set of synaptic weights. When the selected feature is large enough (above a threshold), the node is activated, sending a strong signal down the wire. Otherwise, the node is not activated, and no or a very low level signal will be sent.

This simple one-way flow of information from the input to output layer is the main character of a feed-forward ANN. It is the simplest form of all possible ANN architectures. It is also known as the vanilla type of ANN, popularized by [Hastie et al. (2016)] to mean a feed-forward network with a single hidden layer, the most basic of the commonly used neural network types. There are other fancier types of ANN in which the flow of information are more complicated, such as the convolutional neural network (CNN), and recurrent neural network (RNN). For simplicity, in this book we will discuss only the plain vanilla type of ANN (referred to as VNN from here on, but not necessary with just a single hidden layer).

By combining all the signals from the hidden layer, the output node performs the final calculation and arrives at a numeric answer. During the learning (or training) phase, this answer (or prediction from the ANN) is compared with the observed data. The difference between the prediction and the ground-truth is the error produced by the ANN, due to errors in its synaptic weights. The conventional way to arrive at a correct set of synaptic weights is by back-propagating the error from the output node to the hidden layer and fine-tuning the synaptic weights to reduce the total error. With more and more training data passing through the ANN, the iteration on the synaptic weights converges to an optimal set of values. At that point, the training is considered as accomplished and finally the ANN is ready for deployment.

In some sense, the ANN model works like a nonlinear regression machine that optimizes a set of weighting parameters (i.e. synaptic weights) using a sample of input-output pairs for training. This version of machine learning requires the use of training data, hence it is categorized as supervised learning. The objective function, in this case, is the sum of squared-errors between the prediction and the actual ground-truth from all the input-output pairs, similar to the examples we have seen in previous chapters. This realization immediately makes the ACORN algorithm relevant and applicable to Artificial Neural Networks.

5.2 Deep-learning Neural Networks and Optimization in High-dimensional Systems

For very simple applications, one hidden layer is enough to produce the final output. The number of nodes within a hidden layer is the number of key features required to arrive at an answer to the problem. These features can be thought of as abstractions (or concepts) that help us make sense out of the seemingly incomprehensible variations within the raw data, like picking out the color, shape and surface texture in a collection of millions of pixels from a picture of an apple.

For more complex applications, such as recognizing the face of a human in an image, more hidden layers are required. Simple features such as the shape of the boundaries of different facial parts are extracted in the first layer, then more abstract features, such as the eyes, nose and mouth, are extracted after that. Higher level features, such as determining the gender or age from a human face, required even more hidden layers to accomplish.

The more detail classification is demanded in an image recognition application, the more hidden layers are needed, reflecting the buildup process from simple concepts to more abstract ones in a hierarchical way. This type of architecture of ANN with multi-hidden-layers was initially deployed to increase the accuracy in computer vision and voice recognition software, then eventually has found applications in many different areas, including the famous *AlphaGo*, the AI program that defeated human grandmaster in the ancient Chinese board game of Go.

An ANN with more than one hidden layer is traditionally known as a multilayer perceptron or MLP. A MLP with a large number of hidden layers is called a deep-learning neural network (DNN).

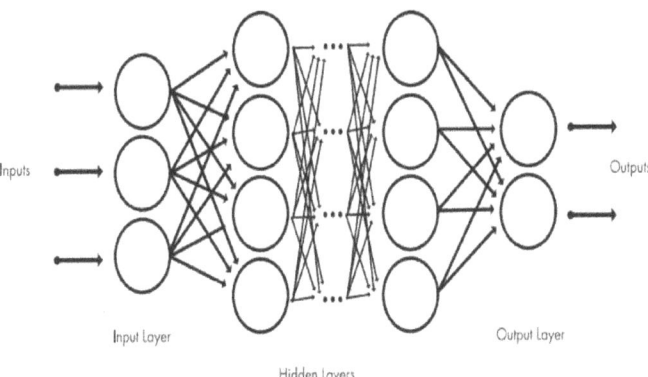

Fig. 5.3: An example of deep-learning neural network (DNN), where the number of hidden layers can be much larger than 1.

In a perceptron with N_i inputs, N_o outputs, and N_n nodes in the (single) hidden layer, the total number of synaptic weights, N_w, between the input and the hidden layer, is

$$N_w = (N_i + 1) \times (N_n) \tag{5.1a}$$

where the off-sets (i.e. the thresholds) have been included (as the $+1$ in the expression above), and the total number of parameters to be optimized in the training process is

$$N_p = (N_i + 1) \times N_n + (N_n + 1) \times N_o. \tag{5.1b}$$

In a DNN, with N_i inputs, N_o outputs, N_h hidden layers and N_n nodes in each hidden layer, the total number of parameters to be optimized in the training process is

$$\begin{aligned} N_p &= (N_i + 1) \times N_n + (N_h - 1) \times (N_n + 1) \times N_n + (N_n + 1) \times N_o \\ &= (N_i + 1) \times N_n + (N_n + 1) \times (N_o + N_n \times (N_h - 1)). \end{aligned} \tag{5.2}$$

For a very modest DNN with $N_i = 2$, $N_o = 1$, $N_n = 3$ and $N_h = 3$, the total number of parameters to be optimized in the training process is 21, a considerably high dimensionality compared with those encountered in ordinary type of problem. In the extreme case of *AlphaGo*, two DNNs with 13-hidden-layers that consist of millions of synaptic connections are deployed.

With such high dimensionality, it is no wonder that the use of DNN has not become trendy until recently when the kind of computing power needed becomes more affordable and widely available. The ACORN algorithm discussed in this book can offer a robust solution to the optimization problem in a DNN because: it is more robust than all garden variety steepest descend optimization methods, and most likely converges to the global instead of a local minimum.

More importantly, ACORN Magic (Minimization algorithms gathered in a cloud), a parallel implementation of the ACORN algorithm, can accelerate the rate of convergence, and scalable to the dimension of the problem, since the scan in (r, p) space is independent of each other and so is easily parallelizable.

5.3 Examples of Neural Network with ACORN Learning Algorithm

As our first example, we use a vanilla type of neural network with one hidden layers which contain 2 neurons as a simple example to demonstrate how the ACORN algorithm can be used to train an ANN.

This neural network has two inputs x_1 and x_2 and two outputs y_1 and y_2 (i.e. 2×2×2). There are totally 8 weights in this neural network, hence it is duded NN8 (as shown in Fig. 5.4). The offsets are all fixed at zero.

The activation function of the neural network is sigmoid function: $\sigma(x) = \frac{1}{(1+e^{-x})}$, a conventional one.

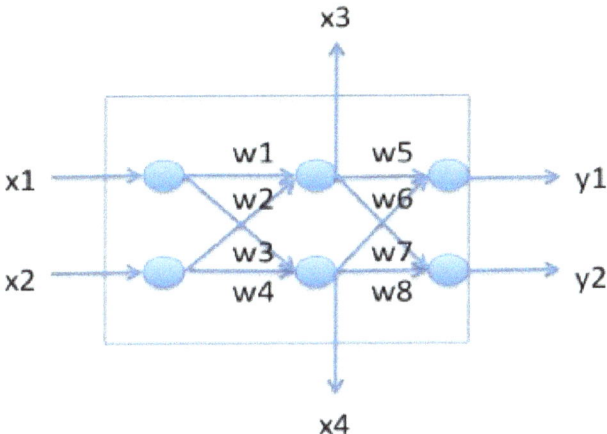

Fig. 5.4:   The configuration of NN8, with weights $w_1, \cdots$ to w_8.

In this example, the neural network weights are simply preset as

$$\begin{pmatrix} 1 & 0 \\ 0 & 1 \end{pmatrix} \text{ and } \begin{pmatrix} 1 & 0 \\ 0 & 1 \end{pmatrix} \tag{5.3}$$

(i.e. identity matrices), then y_1 is only affected by x_1, and y_2 is only affected by x_2.

Using the above weights, and 1200 pairs of (x_1, x_2) chosen randomly from 10^{-8} to 10, we can generate the corresponding output pairs of $(y_{1\_target}, y_{2\_target})$ from the ANN with the preset weights.

To perform training on this neural network (NN8), we randomly initialized the 8 ANN weights, and then feed in the (x_1, x_2) pairs from above to generate their corresponding $(y_{1\_training}, y_{2\_training})$ pairs, which are certainly different from the $(y_{1\_target}, y_{2\_target})$ pairs since we do not know what are the preset weightings.

If the training process is successful, we should be able to recover the preset weights as in Eq. (5.3). In fact, the training is nothing more than optimizing the following objective function, or the sum square of errors:

$$\sum ((y_{1\_training} - y_{1\_target})^2 + (y_{2\_training} - y_{2\_target})^2), \tag{5.4}$$

where the sum is over all the 1200 training data pairs.

At this point, ACORN is used to iterate the 8 weighting variables to find a set of weights so that the above objective function is minimized.

In the following Fig. 5.5, the objective function [Eq. (5.4)] is shown to be reduced steadily from an initial value of 198786 to 0.0098 after 1781

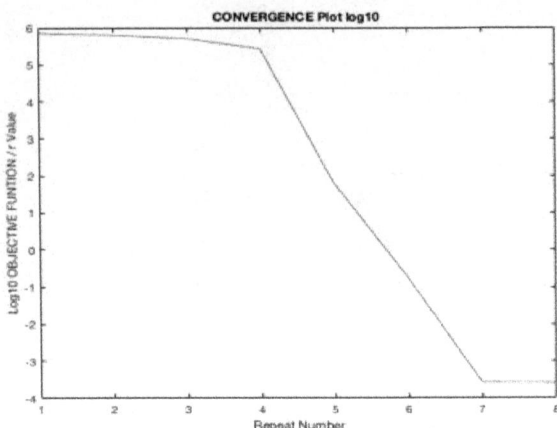

Fig. 5.5:   Convergence of ACORN optimization of NN8.

iterations, with an initial $r = -0.001$ and $p = 12$ in the ACORN algorithm. Notice that the initial objective function value is very large because the neural network contains randomly assigned weights and offsets before it is trained. The number of iterations may vary greatly depending on the maximum number of iteration steps for each r and p values.

We skip the detail of the computation in this example, which is shown in Appendix C for avid readers, in order not to interfere with the flow of our presentation. At the end of the iterating process, the NN8 network weights converges to the preset values given by Eq. (5.2), within an absolute error of 10^{-3}, but a relative error of 10^{-7}. Our training of NN8 using the ACORN algorithm recovered the preset network weightings without a doubt. Even the NN8 neural network discussed here seems to be artificial, however, it is a good example to show how ACORN can effectively get out of local minima and eventually converge to a global minimum. Note that the number of local minima is very large for this example as shown in Fig. 5.6, a surface plot of the objective function vs two variables, w_7 and w_8 (with the other weights kept at their preset values).

It has been long aware[2] that the conventional backpropagation algorithm is subject to the usual problems associated to local minima as it is based on gradient descent. Indeed, many experimenters have found training instances in which the backpropagation process got stuck in such minima.

---

[2]See [Sontag and Sussmann (1989)].

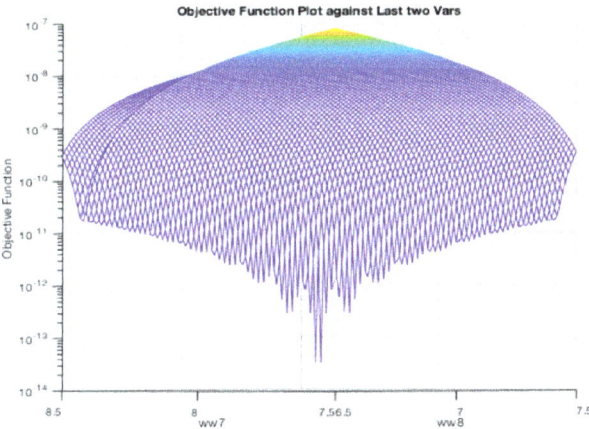

Fig. 5.6: Surface plot of the objective function of NN8 vs two variables, w_7 and w_8 (with the other weights kept at their preset values).

Recently, it has been further shown[3] that for even simple kind of neural network similar to our example here, these local minima are not pathological and the probability of backpropagation training method getting into a local minimum seems to increase towards 1 with the network size.

Our next example is a slightly more complicated $1 \times 20 \times 20 \times 1$ neural network to trace a sine function. In this one-input-and-one-output neural network, which is neither too shallow nor too deep, the number of network parameters to be trained (i.e. optimized) is just 481 (weights and offsets included), according to Eq. (5.1). Hence it is referred to as NN481.

In comparison, a popular and simple VNN [Nielsen (2017)] to recognize hand-written digits from 0 to 9 has an input layer with 784 nodes (to accept an image with 28×28 pixels), one hidden layer with 15 nodes and an output layer with 10 nodes. This amounts to a network with a shocking 11935 parameters (weights and offsets included). It shows that for any real-life applications, the training of ANN is performed in extreme high dimensional space where conventional algorithms are used.

This also explains why even ANN (perceptron) was initially invented by Rosenblatt [Rosenblatt (1958)] in 1958 and created the first wave of enthusiasm for artificial intelligence, it was soon pointed out by Minsky and Papert (1969) [Minsky and Papert (1969)] that algorithms and computers

---

[3]See [Safran & Shamir (2017)].

at that time were not powerful enough to handle the work required by neural networks large enough for any practical purpose. Perceptron researches were abandoned for the next decade or so in a period of time known as the AI winter.

It wasn't until 1986, in a classic paper by Rumelhart, Hinton, and Williams [Rumelhart et al. (1986)], an approach known as backpropagation to perform gradient-descent optimization was demonstrated convincingly to produce fast and effective learning, thus making it possible to use neural nets to solve problems which had previously been eluded solution by ANN. Currently, with the newest powerful computational hardware, such as the GPUs, the backpropagation gradient descent algorithm, as the workhorse of learning in ANN, has achieved huge successes in many real world applications such as speech recognition, computer vision and GoogleGO.

Back to our $1 \times 20 \times 20 \times 1$ VNN (NN481) example, the objective is to train the neural network weights and offsets so that the output of the VNN reproduces sin(x), where x is the input. We choose the input data x, 12,000 data points equally spaced from 0 to 3π, then feed the x data into a sine function, $y = \sin(\text{x})$, to obtain the corresponding 12,000 y_target points. Hence, we get 12000 pairs of (x, y_target) for supervised training purpose. Next, the neural network weights and offsets, all 481 variables, are assigned a random number between 0 to 1. In fact, their range can be flexible. Then we feed the same input data set x into this randomly initialized neural network, NN481, to obtain 12000 y_out values. Obviously, y_out is function of the VNN weights and offsets. The resulting objective function of 481-variables is: $\sum(y\_out - y\_target)^2$. Note that ACORN iteration does not search through the hyper-space of 481 variables. Instead, it uses a two variables search for the convergence control parameters (r, p) to locate the point where the above objective function is minimized, globally.

With 12,000 pairs of training data, it is impossible to minimize the objective function to extremely close to zero. To our best effort, we have found a set of variables (weights and off-sets) so that the objective function is reduced to less than 10^{-3} from an initial value of order unity (normalized at the beginning of the search), after repeating the iterative process 180 times.

This solution was found by using a simple 2-D search procedure to find the best control parameters (r, p) values. The time it took for finding the best solution was about 1 day by using a 900 search points with a single CPU in a current (2018) mid-level consumer grade laptop. Again, the detail of the computation in this example, which is shown in Appendix D for avid readers, is skipped in order not to interfere with the flow of our presentation.

Since the computing time varies with the total number of search points, if the 900 searches can be done in parallel, the computation will be reduced to about 24 min. which is close to the time it takes for backpropagation to find the same solution. Note that the backpropagation method used today has been finely tuned after all the years since its first appearance, while ACORN method has not been tuned yet. The result here confirmed that ACORN is a very useful method to find a warm start solution for backpropagation method.

Another innovative approach to deploy ACORN learning is known as ACORN Magic [Minimization algorithms gathered in a cloud]. In this approach, searches in the 2-D (r, p) space are parceled out in parallel to the servers in a cloud. Each CPU in the cloud performs the algorithm independent of each other. The learning speed of the ACORN Magic approach scales roughly linear with the number of CPUs in the cloud being used. An alternative form of parallelization, GPU computing, may also be able to realize the strength ACORN learning significantly.

In practice, even the training process has finally converged to a stable solution, no one is ever sure that a global minimum is found by the optimization algorithm, be it backpropagation or whatever, due to the number of variables involved is mind-numbing and the objective function is highly non-convex. The usefulness of the DNN so obtained has to be established by a rigorous procedure of validation and testing.

Hence, after the training, the trained NN481 must be validated if it can truthfully reproduce the sin(x) function. Unlike the previous example, NN8, we do not know the exact location of the optimized point in the 481-dimensional space that we searched. All we know is just that the ACORN iteration converged to something that minimized the objective function. As in general, we do not know whether that is a local or global minimum. This is the situation most supervised learning ANN applications end up with. To validate whether our trained NN481 provides a useful/reliable model of the data we have, we reserve part of the available data for validation purpose only. The accuracy of the model is usually measured by the "mean absolute percentage error", or MAPE, of the predicted values forecasted by the model (in this case, the trained NN481) compared with the targeted values computed from sin(x), where x is the set of the validation x data.

For validation purpose, we generated a set of $n = 200$ random numbers, $x_{-validation}$, between 0 and 3π. The corresponding $y_{-output}$ and $y_{-expected}$

are calculated from NN481 and $sin(x_{\_vadidation})$, respectively. Comparing the pairs of $(y_{\_output}, y_{\_expected})$ for the same $x_{\_validation}$, we find that their differences are very small, so that we can conclude that the ACORN trained NN481, to a large extent, recovers the sin(x) function. To be more precise, the MAPE (Mean Absolute Percentage Error), defined in this case as

$$MAPE = \frac{1}{n} \sum_{i=1}^{n} |\frac{(y_{\_output}(i) - y_{\_expected}(i)}{y_{\_expected}(i)}|, \qquad (5.5)$$

is used as a quantitative gauge of the accuracy of the trained NN481.

The result of our simple numerical experiment here gives us a MAPE of 0.1%, which is a pretty good result. In comparison, solving exactly the same problem with backpropagation gives a MAPE of roughly 0.23%.

5.4 Looking Ahead

In an abstract for his talk on "Optimization: Past, Present, Future", in a 2018 Linear Algebra and Optimization Seminar at Stanford University[4], Dr. Robert Bixby, a speaker from Gurobi Optimization Inc, wrote: "For the vast majority of business applications, optimization means linear and mixed-integer programming. Beginning with Dantzig's simplex method in 1947, optimization experienced a slow, uneven period of development into the mid-1980s. Then, beginning in the late 1980s, developments that ensued completely transformed optimization and its applications, driven by truly remarkable performance improvements in the underlying solvers. What's coming next may be even more exciting. Driven by an explosion in available business data, a new broad corporate focus on extracting value from that data, increased computing power, and the continually expanding power of optimization solvers, optimization promises to become an indispensable tool in managing the modern enterprise."

Undoubtedly, optimization has been the engine driving growth of many of the modern technologies such as machine learning and artificial intelligence. In order for ACORN Magic to become one of the next generation truly nonlinear solvers for training DNNs, we will need to come up with a way to implement the procedure currently done manually so that it can be done automatically as well as efficiently.

One solution is to develop and implement an expert system based on the knowledge and experiences from running ACORN magic over a variety of NN training problems as the examples presented in this chapter.

---

[4]https://canvas.stanford.edu/courses/76273

From studying these two examples we believe that manual intervention will always be needed to handle exceptions and to validate the results.

We are ready to take on this important task by the development of a Man-Machine Joint Learning Platform (MJL Platform), a project that is under the umbrella of Big Data Ideal (BDI) Laboratory, a China based company with Stanford heritage.

BDI Lab will provide online courses and training for big-data engineers/technicians/programmers, as well as other big-data related consulting services. For more information, our readers are encouraged to explore the web-pages[5].

Further ahead, we foresee a future of ubiquitous optimization, when autonomous vehicles, appliances, engines, assembly lines, factories ... are all equipped with chips (cpu) hard-wired with ACORN algorithm to fine-tune their operations on-the-fly, sort of "dynamic optimization".

Summary of Chapter 5

The lessons that can be learned from this chapter are

- The training of artificial neural networks is a regression problem to determine the weighting factors that connect the artificial neurons (nodes) of different layers. The number of these weights is often in the millions, especially for deep neural networks. Hence, the amount of data used to train the ANN or DNN is also extremely large. This will lead to an objective function with a large number of local and fictitious minima that will confuse many conventional optimization algorithms.

- In two examples, NN8 and NN481, ACORN algorithm is used to determine the weights in two ANN.

- Due to searching nature of ACORN in (r, p) space, it is conceivable to implement the ACORN algorithm in a parallel computing architecture to accelerate its convergence rate so that solution to more DNN learning problem can be within reach.

---

[5]The link is `http://bigdataideal.com/application/`, which contains more examples in DNN training with ACORN algorithm, performance of ACORN training compared with backpropagation.

Epilogue

Since the first draft of this book, the responses are encouraging. Here are some enthusiastic remarks we received from aficionados of our work.

"Congratulations on publishing the textbook on nonlinear algebra. It is going to transform the way mathematics is taught, from algebra to calculus to topography and multivariate space etc. Great contribution to our knowledge."

"Thank you for the ACORN code. I tried an example included with the code; and it worked impressively fast.... It provides a nice bridge between methodology and applications. During my PhD work I was mostly working on methods (mostly on estimating information criteria in complex settings; and had some experience with optimal design and response surfaces); but after getting a PhD, I was focusing mostly on applied methodology ... looking into ways in which machine learning and network analysis could be used to analyze and catalyze innovation. I am looking forward to trying out ACORN on some of the applications; and try to gain insights from it. I hope I'll be able to make a contribution." ..

The writing of this book has been a big challenge for both of us. We sincerely thank World Scientific for their enthusiasm to publish this work. We hope through the publication of our book, more scientists and engineers will be attracted to the field of nonlinear programming and optimization algorithms. We realize that what we have presented is only the first step in opening a way for practitioners to examine this exciting subject from a new perspective. After working on it for almost 20 years, we think it is time for a book on an adaptive, constraint optimal robust nonlinear minimization algorithm, at least as a simple adjunct to existing algorithms. We wish our readers all the best in using ACORN as an add-on to validate the solutions they have already found. If you need help to get started, please contact us at the web-pages through the link `http://bigdataideal.com/`.

Finally, the second author would like to take this opportunity to thank the first author for the honor and opportunity to participate in this exhilarating book project. Without the years of experience, hard work and tinkering of the first author, this new approach to optimization, ACORN and ACORN Magic, will not be a reality and mature enough to be a viable alternative to mainstream optimization algorithms.

Appendix A

Code for Plotting the Basin of Attraction (BOA) in Octave

```
1   %% To plot Basin of Attraction BOA
2   % Objective function: Rosenbrock function defined by F1, ...
        with two variables
3   %---------------------------------------------------
4   % Warning : On running this the workspace memory will be ...
        deleted. Save if
5   % any data present before running the code !!
6   %---------------------------------------------------
7   % based on:
8   %http://www.mathworks.com/matlabcentral/fileexchange/
9   %        43564-basins-of-attraction?focused=3798465&tab=function
10  % for discussion of algorithms:
11  % https://www.mathworks.com/help/optim/ug/
12  %        constrained-nonlinear-optimization-algorithms.html
13  % https://www.mathworks.com/help/optim/ug/
14  %        first-order-optimality-measure.html#brhkghv-48
15  % The roots of the given governing equations
16  % r1 = [-1 1] ; r2 = [ 1 1] ; r3 = [0 0]; %%no longer needed
17  % This version can count total local minima within the box ...
        %%KT%%13Aug2017
18
19
20  %clear all;
21  %warning('off') % To off the warning which shows "Matrix is ...
        close to singular
22                  % badly scaled" when algorithm passes through ...
                        a point where the Jacobian
23                  % matrix is singular
24
25  F1 = @(xx) xx(1)^2 * (1 - xx(1)^2)^2 + 5*(xx(2) - xx(1)^2)^2 ...
        + 0.01*(xx(2)^2 + xx(1)^2)^2; %function to be optimized
26  F1ab=@(a,b) F1([a b]);    %for plotting using arrayfun
27
```

53

```matlab
28  % setup 2D grid
29  xl=-2;xu=2;yl=-1;yu=3; nxy=160;
30  x = linspace(xl,xu,nxy) ;
31  y = linspace(yl,yu,nxy) ;
32
33  Lplot= 0;    %1 or 0; turn on/off function plot
34  % Plot the function -- just to see what it is.
35  if Lplot
36  [X,Y] = meshgrid(x,y);      %[aa, bb] = meshgrid(x, y);
37  Z= arrayfun(Flab, X, Y);
38  %figure; contour(X,Y,Z,'ShowText','on');
39  v=[0.02,0.1,1,2,4,10,16,25,36,50,80];
40  figure; contour(X,Y,Z,v,'ShowText','on'); xlabel('x'); ...
        ylabel('y');
41  title('Contour Plot x^2*(1 - x^2)^2 + 5*(y - x^2)^2 + ...
        0.01*(x^2 + y^2)^2');
42  %disp('This is our function. Hit any key to proceed with ...
        optimization')
43  %pause
44
45  figure; mesh(x, y, arrayfun(Flab, X, Y));
46    xlabel('x'); ylabel('y'); title('3D Plot x^2*(1 - x^2)^2 + ...
        5*(y - x^2)^2 + 0.01*(x^2 + y^2)^2');
47  disp('This is our function. Hit any key to proceed with ...
        optimization')
48  pause
49  end   %Lplot
50
51  % Initialize the required matrices
52  Xr1 = [] ; Xr2 = [] ; Xr3 = [] ; Xr4 = [];
53  xs1=zeros(100,100); xs2=zeros(100,100); fs=zeros(100,100); ...
        XR=zeros(100,100);
54  xrt1=[]; xrt2=[]; icnt=0;
55  A = []; B = []; Aeq = []; Beq = []; nonlcon = [];
56  options = ...
        optimoptions('fmincon','Display','notify-detailed', ...
        'Algorithm','sqp');
57  %optimoptions('fmincon','Display','iter'); ,'Algorithm','sqp ...
        trust-region-reflective' interior-point
58  disp('Algorithm: sqp')
59  lb = [xl  yl]; ub = [xu  yu];
60  tic
61  for i = 1:length(x)
62      for j = 1:length(y)
63          X0 = [x(i);y(j)];
64          icnt=icnt+1;
65          % Solve the system of Equations using Newton's Method
66          % X = NewtonRaphson(X0) ;
67
68  %KTuse fmincon to optimize %KT
```

```
69    b0 = [x(i) y(j)];
70    [xs, fval] = fmincon(F1,b0,A,B,Aeq,Beq,lb,ub, nonlcon,options);
71    %endKTuse fmincon to optimize %KT
72    xs1(i,j)=xs(1); xs2(i,j)=xs(2); fs(i,j)=fval;
73                if icnt==1
74                 xrt1=[xs(1) xrt1]; xrt2=[xs(2) xrt2]; XR(i,j)=1;
75                else
76                 for k=1:length(xrt1)
77                  ibr=0;
78                  if abs(xs(1)-xrt1(k)) + abs(xs(2)-xrt2(k)) <4e-5
79                   XR(i,j)=k; ibr=1; break;   %XR stored the root#
80                  end
81                 end %k
82                 if ibr==0
83                  XR(i,j)=length(xrt1)+1; xrt1=[xs(1) xrt1]; ...
                        xrt2=[xs(2) xrt2];
84                 end %if
85                end %if
86
87                % Locating the initial conditions according to error
88                %if abs(xs(1)-r1(1)) + abs(xs(2)-r1(2)) <1e-4
89                %    Xr1 = [X0 Xr1]  ;
90                %elseif abs(xs(1)-r2(1)) + abs(xs(2)-r2(2)) <1e-4
91                %    Xr2 = [X0 Xr2] ;
92                %elseif abs(xs(1)-r3(1)) + abs(xs(2)-r3(2)) <1e-4
93                %    Xr3 = [X0 Xr3] ;
94                %else             % if not close to any of the roots
95                %    Xr4 = [X0 Xr4] ;
96                %end
97
98         end %j
99    end %i
100   toc
101   disp('Loop ends. Hit any key to proceed with optimization')
102   pause
103   %warning('on') % Remove the warning off constraint
104
105   for i = 1:length(x)
106        for j = 1:length(y)
107          X0 = [x(i);y(j)];
108          if XR(i,j)==1
109           Xr1 = [X0 Xr1];
110          elseif XR(i,j)==2
111           Xr2 = [X0 Xr2];
112          elseif XR(i,j)==3
113           Xr3 = [X0 Xr3];
114          else              % if none of the above
115           Xr4 = [X0 Xr4];
116          end %if
117        end %j
```

```
118  end %i
119
120  figure
121  set(gcf,'color','w')
122  hold on
123  plot(Xr1(1,:),Xr1(2,:),'.','color','r') ;
124  plot(Xr2(1,:),Xr2(2,:),'.','color','g') ;
125  plot(Xr3(1,:),Xr3(2,:),'.','color','b') ;
126  %plot(Xr4(1,:),Xr4(2,:),'.','color','k') ;
127  title('BOA x^2(1 - x^2)^2 + 5(y - x^2)^2 + 0.01(x^2 + y^2)^2 ...
          fmincon sqp algorithm');
128  xlabel('x'); ylabel('y');
```

Appendix B

ACORN Code Implementation in Octave

We chose to implement the ACORN algorithm in GNU Octave, which is software featuring a high-level programming language, primarily intended for numerical computations, for 3 reasons:

(1) Octave is mature scientific computation language with a user-friendly developmental environment;
(2) Octave is part of the GNU Project. It is public domain free software under the terms of the GNU General Public License, available to anyone interested;
(3) The syntax of Octave is almost 100% compatible with another popular scientific computational platform, MATLAB, which is non-public-domain software but has a large community of devoted users.

Because of the last reason, the codes exhibited below can also run under MATLAB. The ACORN code package contains the following working files: **Aimigo8.m**, **Leesolver.m**, **ModelEqn.m**, **calcfJ.m**, **writeOutputF.m** and **isOctave.m**.

The main file in this group is '**Aimigo8.m**', which is to be called from a driver file outside. The rest of the files are called from within **Aimigo8**, where the main flow of the calculation begins from **Aimigo8** to **Leesolver**, to **ModelEqn**, and finally to **calcfJ**. The other two routines, **writeOutputF** and **isOctave**, are just for book-keeping purposes.

We include here an example driver program (or script in the terminology of Octave/MATLAB) '**fitSinCos.m**'.

Our readers can use this as a template to solve their own problems by inserting their own objective functions in the code below.

B.1   Listing of Driver Code for ACORN

```
 1  %%%%%%%%%%%%%%%begin file for driver:  ...
        fitSinCos.m%%%%%%%%%%%%%%%%%%
 2  % MINA8 is for fast search learning the rules
 3  %==========Read me Guide for Open Source  June 28 2016 by ...
        Martin J. Lee
 4  %==========Revised & commented by Ken Tsang, July 25, 2016
 5
 6  %==========This is a 2D example of finding the best ...
        variables (a, b) to minimize the object function FitFuncAb:
 7  % % Fitting example for function Fab:
 8  %    Fab(a,b,x) = a*cos(b*x) + b*sin(a*x)
 9  % where:
10  %    - a and b are the 2 parameters to be determined by the ...
        minization,
11  %    - in the example we fixed the answer to be something ...
        pre-determined, e.g. [a, b] = wwd = [100.3, 102.3],
12  %    - an vector, g_y, of 101 equidistant points between -50 ...
        and +50,
13  %    - compute corresponding vector of function values g_Ibar ...
        = {Fab(wwd(1), wwd(2), g_y)},
14  %    - now will try to determine the original values of a and ...
        b via curve fitting, i.e.
15  %       vary parameters a and b to minimize the residual of ...
        sum{(Fab(a,b,g_y)} - g_Ibar)^2}
16  % Upon successful fit the {a,b} should become wwd and the ...
        residual will be zero.
17  % parameters a and b are stored in parameter vector ww in ...
        the code.
18
19  clear
20
21  Fab = @(a,b,x) a*cos(b.*x)+b*sin(a.*x); %function to be fitted
22
23  %The boundaries for the variables are defined as (ww0-dw, ...
        ww0+dw)
24  ww0=[100 102];   %input to Aimigo8
25  dw = [0.3 0.3]; %input to Aimigo8
26  wwd=[100.3 102.3]; dw = [0.5 0.5];  % sometimes you need to ...
        Adjust bounds to find solution  % wwd is the desired ...
        solution
27
28  % w0 is the starting point i.e. initial guess
29  w0 = [100 102];   %input to Aimigo8
30  w0 = [-0.00001  -0.0001];
31
```

```
32   g_y = [-50:1:50];      %x values where the function Fab is to ...
         be fitted; in applications 'x' are from your data, which ...
         is known.
33   g_Ibar = Fab(wwd(1), wwd(2), g_y);    %in applications ...
         'g_Ibar' are from your data, which is known, here we ...
         calculate it from Fab
34   FitFuncAb = @(a,b) sum((Fab(a,b,g_y)-g_Ibar).^2);   %this is ...
         the function to be minimized to determine (a,b)
35   FitFunc = @(ww) FitFuncAb(ww(1),ww(2));
36
37
38   if true
39     % Plot the function -- just to see what it is.
40     ta = linspace(ww0(1)-dw(1), ww0(1)+dw(1), 81)';
41     tb = linspace(ww0(2)-dw(2), ww0(2)+dw(2), 81)';
42     [aa, bb] = meshgrid(ta, tb);
43     mesh(ta, tb, arrayfun(FitFuncAb, aa, bb));
44   end
45   disp('This is our function. Hit any key to proceed with ...
         optimization')
46   pause
47
48
49   % % % ===============
50   aimigo_param = {};
51
52   aimigo_param.caseid = 2;
53   aimigo_param.iimax = 50;   % number of iterations
54   aimigo_param.rrstart = -0.001;% r start
55   aimigo_param.ncase = 100;   % number of (p times r) points
56
57   aimigo_param.p0 = 20 % p start
58   aimigo_param.dp = -0.5;% p step size
59   aimigo_param.dr = 0.1;% r step size
60   aimigo_param.dw_p = 200;% p search range
61   aimigo_param.dw_r = 2;% r search range
62
63   % % % ===============fast test
64   aimigo_param.caseid = 2;
65   aimigo_param.iimax = 50;   % number of iterations
66   aimigo_param.rrstart = -0.001;% r start
67   aimigo_param.ncase = 9;   % number of (p times r) points
68
69   aimigo_param.p0 = 3 % p start
70   aimigo_param.dp = -1;% p step size
71   aimigo_param.dr = 0.01;% r step size
72   aimigo_param.dw_p = 200;% p search range
73   aimigo_param.dw_r = 2;% r search range
74
75   aimigo_param.ww0_r = aimigo_param.rrstart;
```

```
76  aimigo_param.ww0_p = aimigo_param.p0 ;
77
78  aimigo_param.w0_p = -0.00001;
79  aimigo_param.w0_r = -0.0001;
80
81
82  obj00=FitFunc(w0);
83  [best_obj, best_vars, num_evals] = Aimigo8(FitFunc, ww0, dw, ...
        w0, aimigo_param);
84
85  bestobj1 = FitFunc(best_vars);
86  fprintf(' Best solution found after %d function evaluations: ...
        %g\n', num_evals, bestobj1);
87  %pause
88  repeat = 1;
89
90  while repeat<19 && bestobj1>1.0e-5 && bestobj1/obj00>1.0e-6
91  ww0 = best_vars
92  [best_obj, best_vars, num_evals] = Aimigo8(FitFunc, ww0, dw, ...
        w0, aimigo_param);
93  bestobj1 = FitFunc(best_vars);
94  fprintf('repeat = %d -- To quit,  hit ^c twice %d function ...
        evaluations: %g\n', repeat, num_evals, bestobj1);
95  %pause
96  repeat=repeat+1;
97  end   %while
98
99  obj00
100 bestobj1
101 best_vars
102 bestobj1/obj00
103 %%%%%%%%%%%%%%end file for driver:  ...
        fitSinCos.m%%%%%%%%%%%%%%%%%%
```

Appendix C

Details for Optimizing the Objective Function in NN8 (Chapter 5)

In this Appendix and the next (Appendix D), we show the intermediate results in running the ACORN package to train the two neural networks discussed in Chapter 5, NN8 and NN481. In spite of the fact that they are very different problems, the ACORN processes are the same and produce similar results. The version of ACORN used is the version without derivative so that the computation time is a lot shorter.

The following graphs are results from running ACORN to train the NN8 neural network discussed in Chapter 5.

At the beginning of the ACORN runs, a 3×3 grid is used for simplicity, so that there are just 9 points in the search grid covering the (r, p) space with an initial guess in the center, and there are fewer lines in the graphs comparing with those in Appendix D. The reason for using such a coarse grid is that each grid-scan can be completed quickly, even we may have to repeat the scanning process more.

For each (r, p) pair in the scan, the ACORN process was carried out to 200 iterations. Out of these 9 runs, one of them achieved the largest reduction in the objective function as shown in Fig. C.1. The convergence of this pair of (r, p) values, which is automatically picked by the ACORN algorithm, was shown in Fig. C.2. Throughout this Appendix and the next, the values of the objective function are all normalized by its initial values in the iteration, so that all traces of the relative objective function are started from 1. The term "objective function" referred to below is abbreviation of "relative objective function", for the sake of convenience.

We repeated the scan with the center of the grid moved to this point in (r, p) space, the lowest objective function was further reduced by the ACORN iteration process.

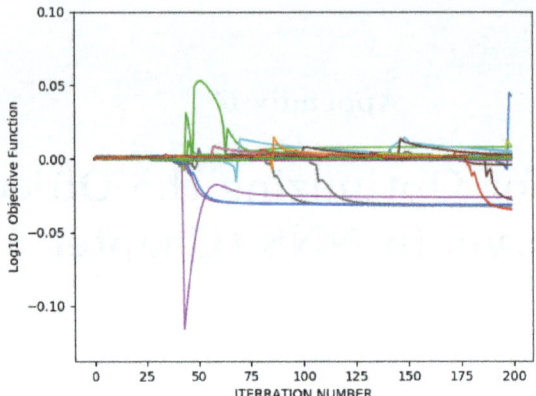

Fig. C.1: The changes in logarithm (base 10) of the relative objective functions as function of iteration, starting from the same initial guess but with different (r, p) pair in the scan.

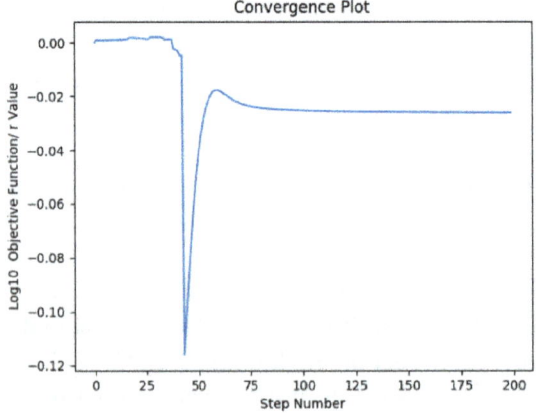

Fig. C.2: The change in logarithm (base 10) of the objective function, which achieved the largest reduction in the objective function out of all (r, p) pairs in the scan, is shown as function of iteration.

Notice in some runs, as shown in Fig. C.3, the lines are very flat. These flat lines indicate that starting point is inside of the basin of attraction of a minimum objective function.

Repeating the (r, p) scan, ACORN may get out of the basin and get into the basin of attraction of another local minimum solution.

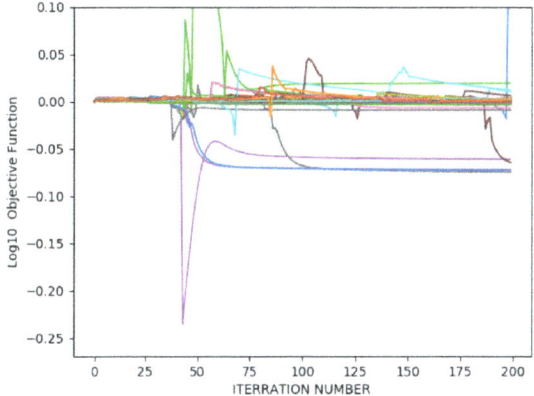

Fig. C.3:   Repeat the scan with center of the grid moved to the (r, p) pair that achieved lowest value in the objective function in the previous one.

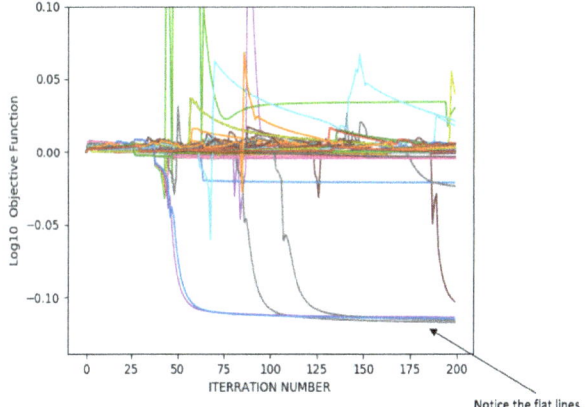

Fig. C.4:   Repeat the scan with center of the grid moved to the (r, p) pair that achieved lowest value in the objective function in the previous one (Fig. C.3).

In some large basins, repeating running ACORN may not make the objective function decrease very fast. Notice the scale in the plot that followed, the objective function is not decreasing as fast as previous plots.

Sometimes, ACORN may seem to be trapped. The objective function is not decreasing much at all.

If we change some ACORN search parameters such as the number of search points or the range of the search domain, ACORN is able to get out of the basin. The objective function is decreasing again as shown in Fig. C.8.

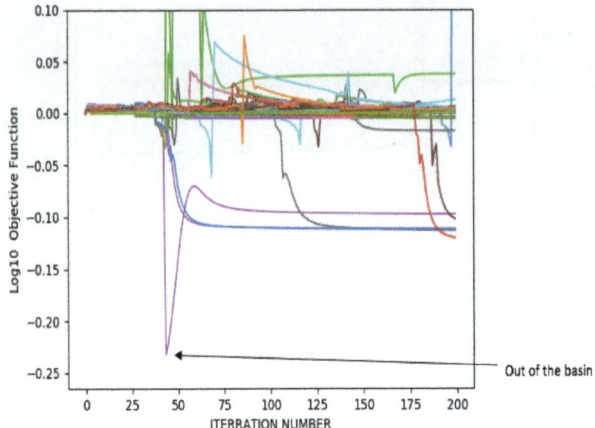

Fig. C.5: Repeat the scan with center of the grid moved to the (r, p) pair that achieved lowest value in the objective function in the previous one (Fig. C.4).

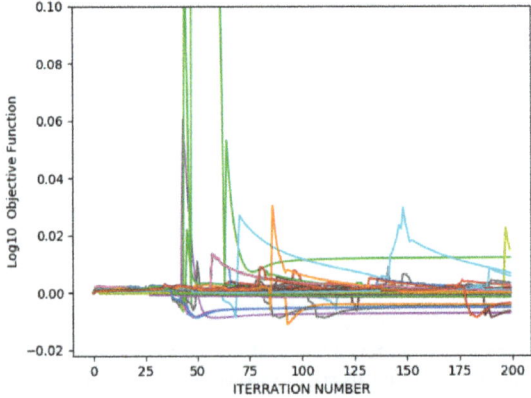

Fig. C.6: Repeat the scan with center of the grid moved to the (r, p) pair that achieved lowest value in the objective function in the previous one (Fig. C.5).

Now repeat running ACORN, the search for minimum of objective function can continue.

There are many local minima in the NN8 neural network. We need to keep tuning the ACORN internal parameters to avoid getting stuck.

After changing the ACORN parameters, the objective function will decrease faster again.

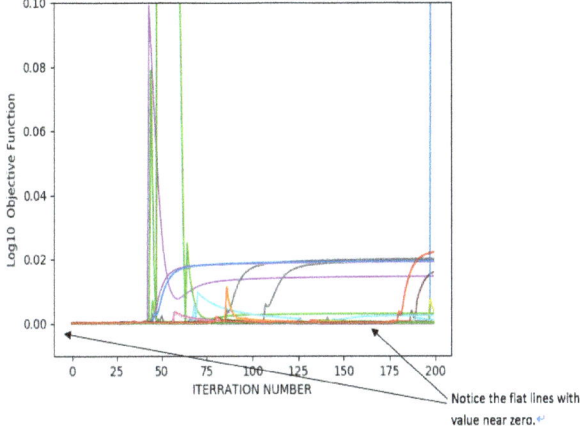

Fig. C.7:   Repeat the scan with center of the grid moved to the (r, p) pair that achieved lowest value in the objective function in the previous one (Fig. C.6). At this point it seems the search is trapped in a local minimum.

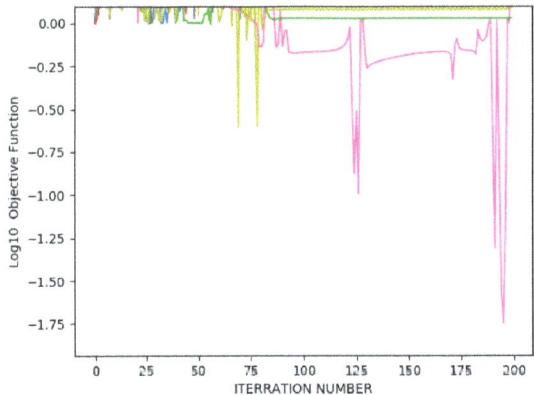

Fig. C.8:   Repeating the scan by changing the range of the search domain in the previous one (Fig. C.7), the objective function resumed its downward move.

When the objective function is getting closer to the global minimum, the objective function may not decrease as fast as the beginning searches.

Sometimes, we can catch a big break. The objective function will decrease very fast again.

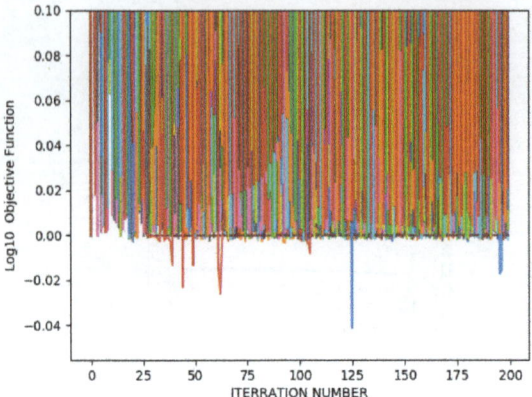

Fig. C.9: Repeat the scan with center of the grid moved to the (r, p) pair that achieved lowest value in the objective function in the previous one (Fig. C.8). Since there are many local minima in the NN8 neural network, we need to keep refining the mesh in the (r, p) scan to avoid get trapped again.

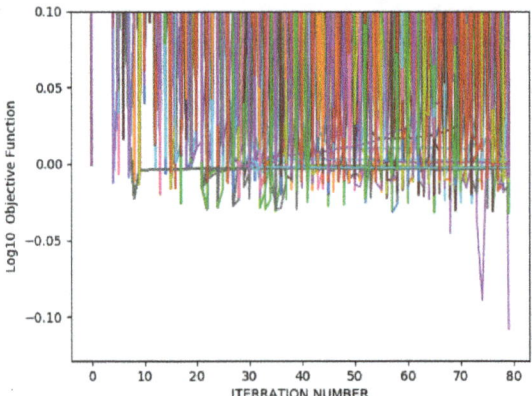

Fig. C.10: Repeat the scan process several times with center of the grid moved to the (r, p) pair that achieved lowest value in the objective function in the previous one. Occasionally, if it seems the search is trapped in a local minimum, we need to enlarge the domain of search and refine the mesh in the (r, p) scan to get out of the trapped region.

The plots above can give the users some insights into how the objective function behaves as ACORN searches for the global minimum for a simple ANN example with many local minima.

Appendix D

Details for Optimizing the Objective Function in NN481 (Chapter 5)

The following graphs are obtained from running ACORN to train the NN481 neural network discussed in Chapter 5.

Running ACORN with randomly initialized weights and offsets and with $r = 0.001$ and $p = 12$ in a 10×10 grid (100 search points) yields the following plots. A finer and larger grid is used in this case because this is a search in a higher dimensional space.

Throughout this Appendix, for convenience, the values of the objective function are all normalized by its initial values in the iteration, so that all traces of the relative objective function are started from 1. The term "objective function" referred to below is the abbreviation of "relative objective function".

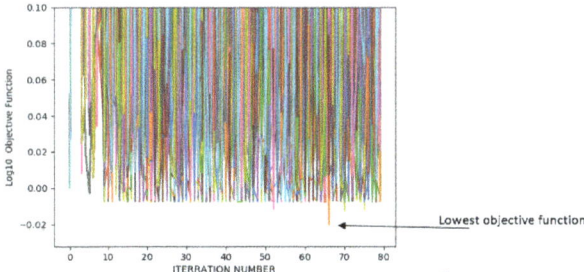

Fig. D.1: The changes in logarithm (base 10) of the relative objective functions as function of iteration, starting from the same initial guess but with different (r, p) pair in the scan, which is a finer grid over a larger domain in (r, p).

Each color in the plot represents a pair of r and p values. Among all the different r and p pairs, there is one pair, picked up by ACORN algorithm automatically, producing the lowest objective function.

With the (r, p) pair that produces the lowest objective function in Fig. D.1, we can plot the changes in objective function step by step. From the Convergence Plot in Fig. D.2, we can see how the objective function varies in each iteration step.

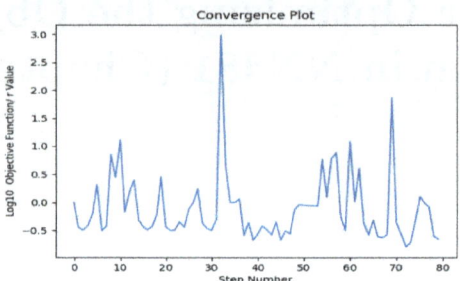

Fig. D.2:   The change in logarithm (base 10) of the objective function, which achieved the largest reduction in the objective function out of all (r, p) pairs in the scan, is shown as function of iteration steps.

We repeat the (r, p) grid scan again with the center of the grid moved to the (r, p) values that give us the smallest objective function and based on the weights and offsets obtained from the above (r, p) pair. A similar plot is shown in Fig. D.3 the set of objective functions achieved in this new scan.

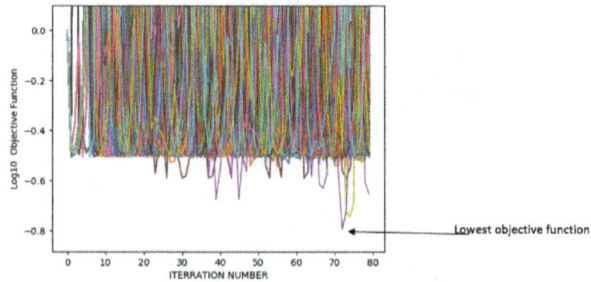

Fig. D.3:   Repeat the scan with center of the grid moved to the (r, p) pair that achieved the lowest value in the objective function in the previous one (Fig. D.2).

Run ACORN again, the objective function becomes smaller. Notice the log value of the objective function is negative, which shows that the objective function is smaller than the previous one.

The objective function does not decrease the same amount in every repeated run. In some runs, the objective function decreases less, as in Fig. D.3.

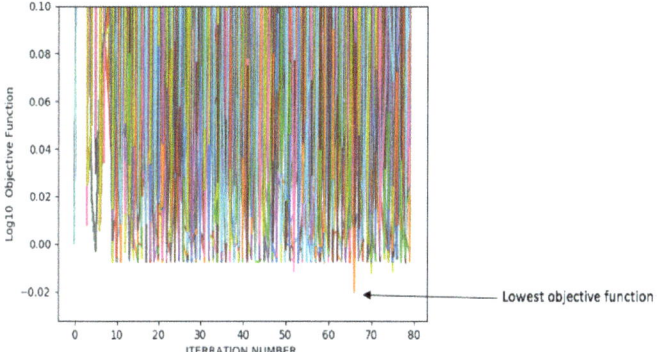

Fig. D.4:   Repeat the scan with center of the grid moved to the (r, p) pair that achieved the lowest value in the objective function in the previous one (Fig. D.3).

Some runs, the objective function decreases more, as in Fig. D.4.

With every repeat, the objective function gets smaller and smaller.

Finally, when the accuracy requirement is achieved, the repetition in (r, p) scan is stopped.

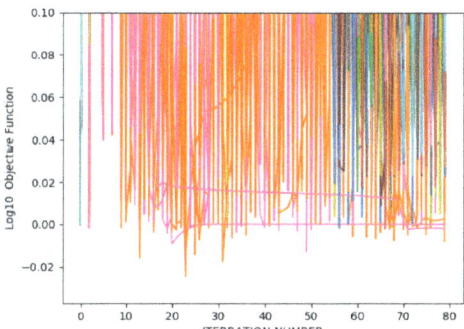

Fig. D.5:   Repeat the scan a few times with center of the grid moved to the (r, p) pair that achieved the lowest value in the objective function in the previous one. Finally, stop when the desired accuracy is achieved.

Bibliography

Marquardt, D. W. (1963). An Algorithm for Least-Squares Estimation of Nonlinear Parameters, *SIAM Journal on Applied Mathematics*, 11 (2), 431-441. doi:10.1137/0111030

Sontag, E. D. and Sussmann, H. J. (1989). Backpropagation Can Give Rise to Spurious Local Minima Even for Networks without Hidden Layers, *Complex Systems*, 3, 91-106.

Safran, I. and Shamir, O. (2017). Spurious Local Minima Are Common in Two-Layer ReLU Neural Networks, *arXiv Technical Report*.

Levenberg, K. (1944). A Method for the Solution of Certain Non-Linear Problems in Least Squares, *Quarterly of Applied Mathematics*, 11 (2), 431-441. doi:10.1137/0111030

Lee, M. J. (2012). IMIGO: An Optimal Adaptive Nonlinear Program for Accelerator Optics Modeling and Other Applications, Stanford Linear Accelerator Center.

Nielsen, M. (2017). Using Neural Nets to Recognize Handwritten Digits, `http://neuralnetworksanddeeplearning.com/chap1.html`

Minsky, M. and Papert, S. (1969). *Perceptions: An Introduction to Computational Geometry*, Cambridge, MA: MIT Press.

Myers, R. H. (1990). *Classical and Modern Regression with Applications* (2nd ed.), Duxbury Press.

Nocedal, J. and Wright, S. J. (2006). *Numerical Optimization* (2nd ed.), Springer Series in Operations Research, Berlin: Springer Verlag.

Rosenblatt, F. (1958). The Perceptron: A Probabilistic Model for Information Storage and Organization in the Brain, *Psychological Review*, 65 (6), 386-408.

Rumelhart, D. E., Hinton, G. E. and Williams, R. J. (1986). Learning Representations by Back-propagating Errors, *Nature*, 323 (6088), 533-536. doi:10.1038/323533a0.

Haykin, S. (2011). *Neural Networks and Learning Machines* (3rd ed.), Pearson.

Hastie, T., Tibshirani, R. and Friedman, J. (2016). *The Elements of Statistical Learning: Data Mining, Inference, and Prediction* (2nd ed.), Berlin: Springer Verlag.